# ELLIPSOMETRY - PRINCIPLES AND TECHNIQUES FOR MATERIALS CHARACTERIZATION

Edited by **Faustino Wahaia**

**Ellipsometry - Principles and Techniques for Materials Characterization**
http://dx.doi.org/10.5772/65558
Edited by Faustino Wahaia

**Contributors**

Eralci Moreira Therézio, Alexandre Marletta, Patricia Targon Campana, Gustavo Dalkiranis, André Vieira, Hugo Gallardo, Ivan Bechtold, Hajime Shitai, Yu Faye Chao, Chien-Yuan Han, Yu-Faye Chao, Yasuaki Hayashi, Akio Sanpei, Jingtao Zhu, Mingqi Cui, XiaoLong Zhou, Saitoh Hidetoshi

**Published by InTech**

Janeza Trdine 9, 51000 Rijeka, Croatia



**Publishing Process Manager** Romina Rovan
**Technical Editor** SPi Global
**Cover** InTech Design team

First published November, 2017
Printed in Croatia
Legal deposit, Croatia: National and University Library in Zagreb

Additional hard copies can be obtained from orders@intechopen.com

Ellipsometry - Principles and Techniques for Materials Characterization, Edited by Faustino Wahaia
p. cm.
Print ISBN 978-953-51-3623-1
Online ISBN 978-953-51-3624-8

For EU product safety concerns:
IN TECH d.o.o., Prolaz Marije Krucifikse Kozulić 3, 51000 Rijeka, Croatia,
info@intechopen.com or visit our website at intechopen.com.

# Meet the editor

Faustino Wahaia received his BS and MSc degrees from the University of Sofia St. Kliment Ohridski, Sofia, Bulgaria, and PhD degree in Physics from the University of Porto, Portugal. He did his postdoctoral research focusing on activity in generation, detection, and characterization of terahertz (THz) waves using photoconductive antennas (PCA) at the Institute of Biomedical Engineering (INEB), University of Porto, Portugal, and the State Research Institute of Lithuania Center for Physical Sciences and Technology, Vilnius.
Dr. Wahaia's research interests are lasers, ultrafast lasers and optics, Raman spectroscopy, imaging and spectroscopic ellipsometry, infrared and THz photonics for diagnostic, ultrafast biophotonics, nanomaterials studies, food analysis, agriculture, defense, health care, homeland security, environment, and pharmacy assessment.
He is presently a researcher at Porto University. He is a member of SPIE, OSA Optical Society, IEEE, and EACR.

# Contents

# Preface

Ellipsometry that measures a change in polarization as light reflects or transmits from or through a medium has been used as nondestructive measurement tool since more than a century. It is a very attractive technology for both academic researchers and industry. The technique has diverse potential applications such as thin-film characterization, ion sensing, surface molecular engineering, and integrated circuit technology among others from a broad list. Up until now, versatile and more reliable ellipsometric systems working at a frequency range from the far-infrared to the ultraviolet of the electromagnetic spectrum for materials studies have been engineered. This dynamic at developing experimental ellipsometric techniques have been enabling I&D researchers with scientific and technical instruments to face the challenges in modern materials science.

This book entitled *Ellipsometry: Principles and Techniques for Materials Characterization* consists of seven chapters with some of the best theoretical reviews on the topic and research works presented hitherto. The chapters cover some ellipsometry application modalities and methods including state-of-the-art and innovative experimental techniques in materials characterization—fundamental approaches on ellipsometry measurements in line with the typical data analysis procedures. Methods such as three-intensity measurement technique in a PSA (polarizer, sample, and analyzer), X-ray diffraction (XRD), Fourier transform infrared spectroscopy (FTIR), scanning electron microscopy (SEM), and classification of diamond-like carbon (DLC) films have been some of the working tools of the authors.

**Dr. Faustino Wahaia**
University of Porto,
Porto, Portugal

# Chapter 1

# Mie-Scattering Ellipsometry

Yasuaki Hayashi and Akio Sanpei

Additional information is available at the end of the chapter

http://dx.doi.org/10.5772/intechopen.70278

**Abstract**

The size and refractive index of particles can be analyzed through the measurement of polarization state of scattered light. The change of polarization state in Mie scattering has been represented by ellipsometric parameters, $\Psi$ and $\Delta$, like the reflection ellipsometry. The analysis method is called Mie-scattering ellipsometry. By in-process Mie-scattering ellipsometry, the growth processes of carbon particles in argon plasma and in methane plasma were analyzed. It was found that carbon particles grow by coagulation in argon plasma, while they grow by carbon coating in methane plasma. It is also shown that imaging Mie-scattering ellipsometry has the potential for the easier confirmation of optical adjustment from a long distance, as well as for the analysis of spatial distribution of particle size.

**Keywords:** Mie-scattering ellipsometry, Mie scattering, fine particles, monitoring, imaging ellipsometry, imaging Mie-scattering ellipsometry, dusty plasma, Coulomb crystal

## 1. Introduction

In the same way as in the commonly-used ellipsometry for thin-film analysis by the reflection of polarized light, the size and refractive index of particles can be analyzed through the measurement of polarization state of the scattered light. The change of polarization state in the size range of Mie scattering is able to be defined by using ellipsometric parameters, $\Psi$ and $\Delta$, like the reflection ellipsometry. The analysis method is called Mie-scattering ellipsometry [1–3].

The evaluation of size and size distribution of fine particles is important for their research and production. Since 1994, it has become possible to observe fine particles stationarily suspended in plasmas forming Coulomb crystals [3–5]. In such states, the position and movement of individual fine particles are easily analyzed.

Mie-scattering ellipsometry, including polarization-sensitive light scattering measurement method, was applied for the analyses of the growth of fine particles of amorphous carbon

[2, 3, 6–9] and amorphous silicon [10]. Lately, image sensors were used for the detection of scattered light instead of photodetector [11, 12]. It is a kind of imaging ellipsometry of Mie scattering. Imaging Mie-scattering ellipsometry has the potential for the analysis of spatial distribution of particle size and the easy confirmation of optical adjustment from a long distance [12].

Mie-scattering ellipsometry in $\Psi$-$\Delta$ two-dimensional (2D) plane enables more detailed analysis than the conventional polarization-ratio method [13] or polarization-sensitive method [9], which effectively applies only one ellipsometric parameter in the size region of monotone function. Although "scattering ellipsometry" was termed in the book of "Ellipsometry and Polarized Light" by Azzam and Bashara [14], the detailed description of its analytical method is not shown there. The analysis of trajectory of ellipsometric parameters depicted in the $\Psi$-$\Delta$ coordinate plane by *in situ* ellipsometry during growth of fine particles brings a lot of information about the growth in the same way as the *in situ* refection ellipsometry during film growth [15–21].

In this chapter, the principium of Mie-scattering ellipsometry, its application, imaging method, and their systems are presented.

## 2. Principium of Mie-scattering ellipsometry

In the same way as in the reflection ellipsometry, ellipsometric parameters $\Psi$ and $\Delta$ for the Mie scattering are defined by the ratio between two complex scattering amplitudes in the direction parallel to the scattering plane and their vertical direction [1]. When particles are spherical and monodisperse, the ellipsometric parameters $\Psi$ and $\Delta$ are defined by the ratio of the scattering amplitude functions [22] of a parallel polarization component ($S_p$), which is in the scattering plane, to that of a perpendicular one ($S_s$) as

$$\tan\Psi \cdot \exp(i\Delta) = \frac{S_p}{S_s} \tag{1}$$

$S_p$ and $S_s$ are complex numbers and functions of the scattering angle, the diameter of particles, and the refractive index. The procedures of calculation of $S_p$ and $S_s$ from particle size and its optical index are given in [22, 23].

When particles are polydisperse in the size range of Mie scattering, the scattered light generally results in some depolarization even if the incident light is fully polarized [24]. In this case, the Stokes vector and the Mueller matrix are required for the calculation of polarization state. When incident light is linearly polarized with the azimuth at 45° to the scattering plane, the Stokes vector of scattered light from a particle "*i*" of a certain diameter is

$$\begin{pmatrix} I_i \\ Q_i \\ U_i \\ V_i \end{pmatrix} = I_0\left(|S_p|^2 + |S_s|^2\right)\begin{pmatrix} 1 \\ -\cos(2\Psi_i) \\ \sin(2\Psi_i)\cos(\Delta_i) \\ -\sin(2\Psi_i)\sin(\Delta_i) \end{pmatrix}, \tag{2}$$

where $\Psi_i$ and $\Delta_i$ are the ellipsometric parameters for the particle " $i$," and $I_0$ is a constant proportional to the particle density and independent of the diameter. The total Stokes vector of polydisperse particles is obtained by the summation of all Stokes vectors as

$$I = \sum_i I_i, \quad Q = \sum_i Q_i, \quad U = \sum_i U_i, \quad V = \sum_i V_i. \tag{3}$$

Generally, the two polarization components of the light scattered from polydisperse particles are imperfectly coherent. When the degree of coherence is expressed by $\mu$, the ellipsometric parameters $\Psi$ and $\Delta$ for the particles are defined here with the Stokes parameters as

$$\begin{pmatrix} I \\ Q \\ U \\ V \end{pmatrix} = I \begin{pmatrix} 1 \\ -\cos(2\Psi) \\ \mu \sin(2\Psi)\cos(\Delta) \\ -\mu\sin(2\Psi)\sin(\Delta) \end{pmatrix}, \tag{4}$$

that is,

$$\Psi = -\frac{1}{2}\cos^{-1}\frac{Q}{I}, \tag{5a}$$

$$\Delta = -\tan^{-1}\frac{V}{U} \quad \text{for } U \geqq 0, \quad \text{otherwise} \bigoplus 180^\circ. \tag{5b}$$

Thus, the ellipsometric parameters $\Psi$ and $\Delta$ for polydisperse particles are obtained independently of the degree of coherence $\mu$.

## 3. Monitoring and analyses by Mie-scattering ellipsometry

### 3.1. $\Psi$-$\Delta$ trajectory in growth of polydisperse fine particles

The ellipsometric parameters of polydisperse particles are calculated as functions of mean size and size distribution spread. For the calculation, the lognormal size distribution function, which was verified as appropriate to apply to the coagulation process of particles [25], is expressed as

$$N(D) = \frac{1}{\sqrt{2\pi} D \ln\sigma} e^{-\frac{(\ln D - \ln D_m)^2}{2(\ln\sigma)^2}}, \tag{6}$$

where $D$ is the particle diameter, $D_m$ is the geometric mean diameter, and $\sigma$ is the geometric standard deviation.

**Figure 1** shows the evolution of the ellipsometric parameters in the increase of geometric mean diameter with the refractive index of 2.5–0.75$i$, which is the value for an evaporated carbon foil, the scattering angle of 90°, and the geometric standard deviations of 1.0 (monodispersion), 1.2, and 1.5.

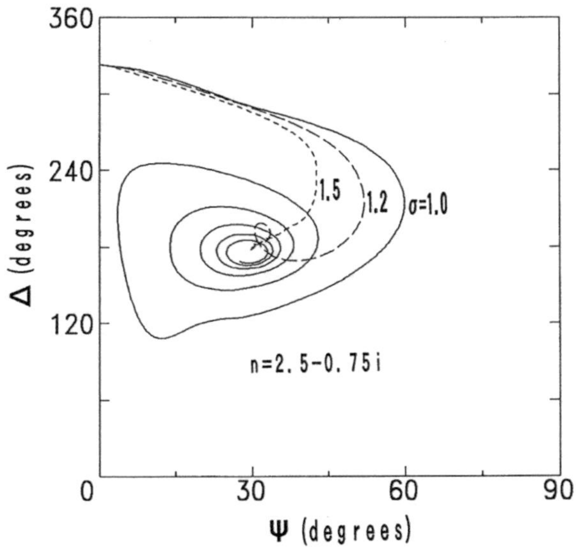


**Figure 1.** Calculated trajectories of the ellipsometric parameters for growing particles. Values used in the calculation are: 2.5–0.75*i* for the refractive index, 90° for the scattering angle, and 1.0, 1.2, and 1.5 for the geometric standard deviations [1].

### 3.2. System layout for Mie-scattering ellipsometry

**Figure 2** shows the system layout for the Mie-scattering ellipsometry. The incidence light beam from an argon-ion laser (wavelength: 488 nm, output power: 100 mW) was linearly polarized at an azimuth of 45° from the plane of scattering through a Glan-Taylor prism polarizer. Its polarization state is expressed as $\chi = 1$, because $E_p = E_s$ . After the scattering from particles in plasma, the polarization state of the light $\chi'$ was determined at a right angle from the incident beam line by use of a rotating-analyzer system directly driven by a stepper motor [26]. Light intensity was measured during the rotation of analyzer by two modes: with and without a quarter-wave plate, by using a computer-controlled in-and-out mechanism. When the fast axis of the quarter-wave plate is set in the scattering plane, scattered light intensity changes with analyzer azimuth $A$ as

$$\mathscr{I}(A) = I + Q\cos 2A + U\sin 2A \quad \text{for without quarter-wave plate,} \tag{7a}$$

$$\mathscr{I}(A) = I + Q\cos 2A + V\sin 2A \quad \text{for with quarter-wave plate.} \tag{7b}$$

Thus, all four Stokes parameters are determined through the Fourier integral of Eqs. (7a) and (7b). Then, the ellipsometric parameters $\Psi$ and $\Delta$, which is uniquely determined in the full range of 360°, are calculated by Eqs. (5a) and (5b).

A parallel-plate rf (radio frequency) discharge reactor (rf electrode diameter: 5 cm, grounded electrode size: 22 cm × 6.5 cm, and separation of the electrodes: 2.0 cm) was used in the experiments (**Figure 3**). Argon or methane gas was introduced into the chamber at a position far from the plasma region so as not to disturb or evacuate the suspended particles with its flow. After the generation of plasma at a pressure of 40 Pa and an rf power of 2 or 8 W, the

http://dx.doi.org/10.5772/intechopen.70278

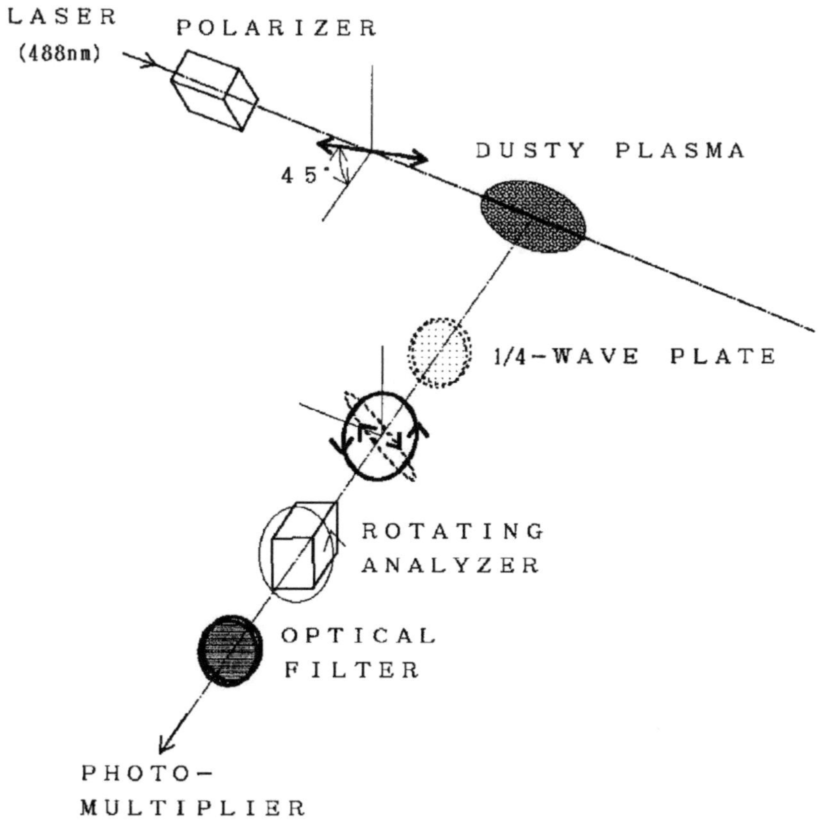


**Figure 2.** Schematic arrangement for the measurement of Mie-scattering ellipsometry. Incident light from an argon-ion laser is linearly polarized. The polarization state of light scattered from particles is analyzed with the use of a quarter-wave plate and a rotating-analyzer system [1].

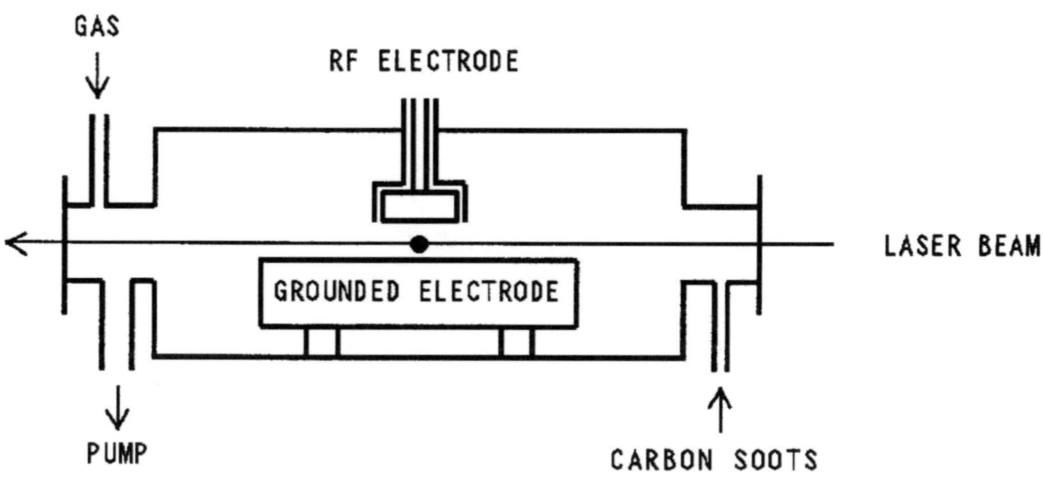


**Figure 3.** Experimental setup for the measurement of carbon particle growth [2].

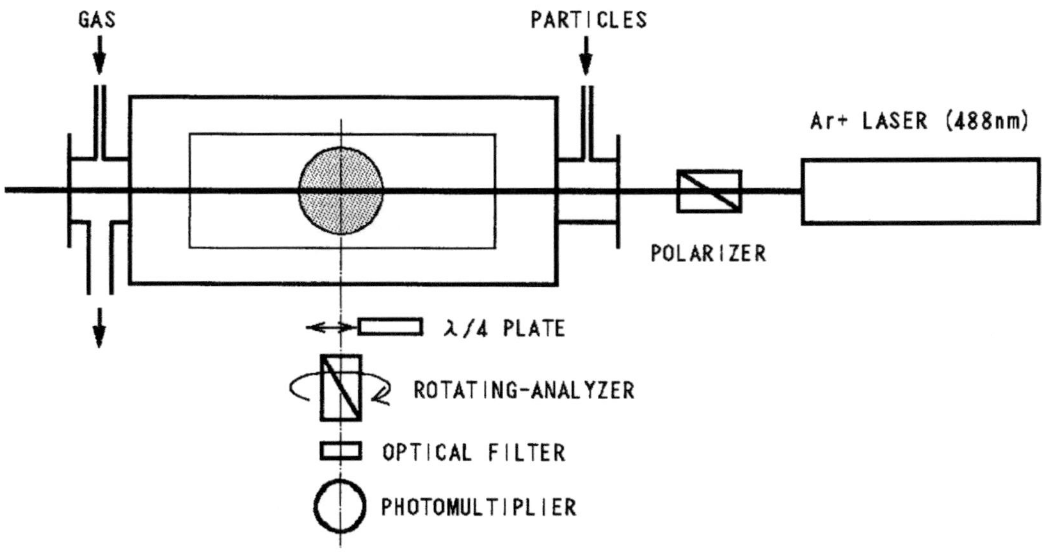


**Figure 4.** System configuration for Mie-scattering ellipsometry measurement in rf plasma reactor (top view).

ultra-fine particles of carbon were supplied all at once into the plasma. Then the ellipsometric measurement was started. The observation point was fixed at a position 1.2 cm below the rf electrode on the center axis, which was located in the plasma bulk.

System configuration for the measurement of Mie-scattering ellipsometry during the growth of carbon fine particles in the rf plasma reactor is shown in **Figure 4**.

### 3.3. *In situ* measurement by Mie-scattering ellipsometry

#### *3.3.1. Carbon particle growth by coagulation in argon plasma*

Argon gas was introduced into the chamber at a position far from the plasma region so as not to disturb or evacuate the suspended particles with its flow. After the generation of argon plasma at a pressure of 40 Pa and an rf power of 2 W, the particles of carbon soot were supplied all at once into the plasma. Then the ellipsometric measurement was started [1].

The particles were observed from the scattering light to be located at both sides of the measurement position on the laser beam line in the early period. At about 700 seconds after the injection, the light intensity gradually increased at the measurement position. The induction period depended on the experimental conditions. The evolution of the ellipsometric parameters after the initial period is shown on the $\Psi$-$\Delta$ coordinate plane in **Figure 5(a)**. A data point shows the average value of five measurements and data were taken every 9 seconds. During the evolution, the parameter $\Delta$ decreased gradually at first and rapidly. The evolution shows larger $\Delta$ at the beginning as compared with the calculated trajectories in **Figure 1**. It is understood from the calculation that the difference depends mainly on the employed refractive index of particles.

In order to fit the calculated trajectory to the experimental data, the simulation was carried out changing parameters; in steps of 0.1 for the real part of the refractive index, 0.05 for the imaginary part, 1 degree for the scattering angle, and 0.1 for the geometric standard deviation along with the assumption of lognormal size distribution of polydisperse particles. The best-fit trajectory was obtained at the refractive index of 2.3–0.35*i*, the scattering angle of 91°, and the standard deviation of 1.5 as shown in **Figure 5(b)**. The smaller value of refractive index than that of evaporated carbon may be due to the inclusion of voids in the carbon particles. It can be estimated from the geometric standard deviation of 1.5 that 68% of all particles have diameters from $D_m/1.5$ to $1.5D_m$.

From the correspondence to the best-fit trajectory, the geometric mean diameter was determined for each experimental data point. Then the particle density was evaluated with the known diameter and the measured scattered intensity. **Figure 6** gives the results and shows that the density gradually increases until the particle begins to grow and then rapidly decreases.

The bright region in the plasma formed by Mie scattering, where particles were distributed, was observed by the naked eye under the same plasma conditions. The region was initially dome-shaped. However, it became concentrated at the center axis and moved toward the rf electrode at the same time [23]. The transition may be due to the time evolution of plasma potential interacting with and interacted by negatively charged particles. The particle transfer and concentration causes coagulation, resulting in the increase of the diameter. Then, the plasma potential distribution is affected and enhances the movement. The abrupt decrease of the density at around 800 s in **Figure 6** supports this speculation.

### *3.3.2. Carbon particle growth by coating in methane plasma*

The inlet port of methane gas was located near a pumping port so as not to disperse particles trapped in methane plasma, which was maintained under the conditions of 40 Pa gas pressure, 16 sccm gas flow rate, and 8 W rf power. Ultra-fine carbon particles were injected instantaneously into the plasma through another gas inlet port [2].

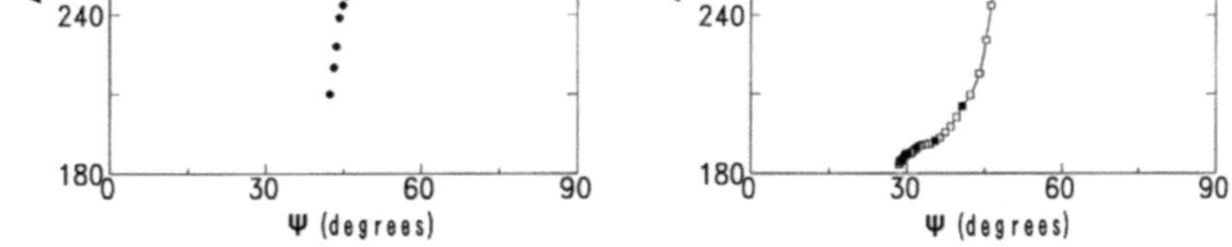


**Figure 5.** Experimental (a) and simulated (b) results of evolution of the ellipsometric parameters. The solid line (in (b)) shows the best-fit simulated trajectory: values used in the calculation are 2.3–0.35*i* for the refractive index and 91° for the scattering angle. The rectangles on the best-fit trajectory indicate geometric mean diameter increases every 20 nm, and the closed rectangles those every 100 nm [1].

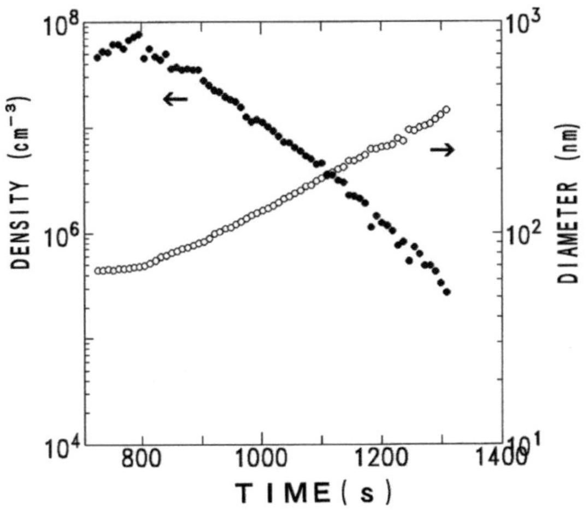


**Figure 6.** Evolution of density and geometric mean diameter of particles in trap at the measurement point in plasma bulk. Closed circles indicate the density and open circles the mean diameter [1].

After the injection of ultra-fine carbon particles in the plasma, the bright region of Mie scattering was observed visually. The region existed only around the sheath-plasma boundary near the rf electrode for the first 300 second, and then extended into the plasma bulk in the shape of a dome, keeping the almost same shape for about 2000 seconds.

The evolution of the ellipsometric parameters during the period is shown by points on the $\Psi$-$\Delta$ coordinate plane in **Figure 7**. The data were taken from 200 to 2300 seconds after the injection of ultra-fine carbon particles in the plasma. First, the parameter $\Psi$ increased gradually with a slight change of $\Delta$, and at $\Psi > 45°$, loops running counterclockwise were observed on the coordinate plane. Similar trajectories of the ellipsometric parameters with the same speed of

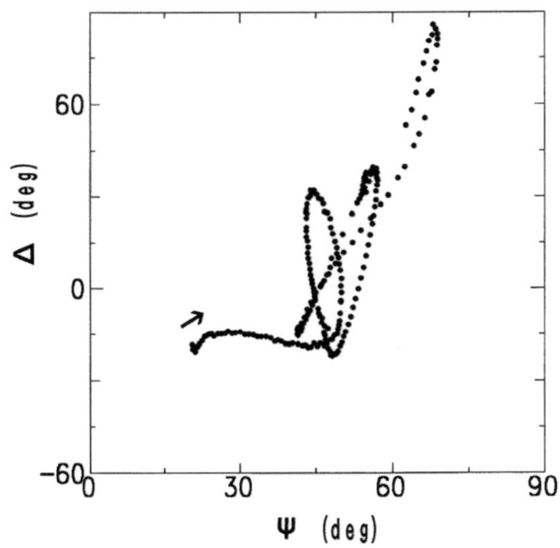


**Figure 7.** Experimental results of evolution of the ellipsometric parameters during carbon particle growth [2].

evolution were depicted at the other measurement positions on the center axis, except for the plasma-sheath boundaries. From the results mentioned above, it can be supposed that the particles in the plasma bulk had the same size distribution during the period.

The calculation of Mie-scattering ellipsometry of spherical polydisperse particles [1] was performed for the analysis of the experimental data. **Figure 8(a)** shows the results of the

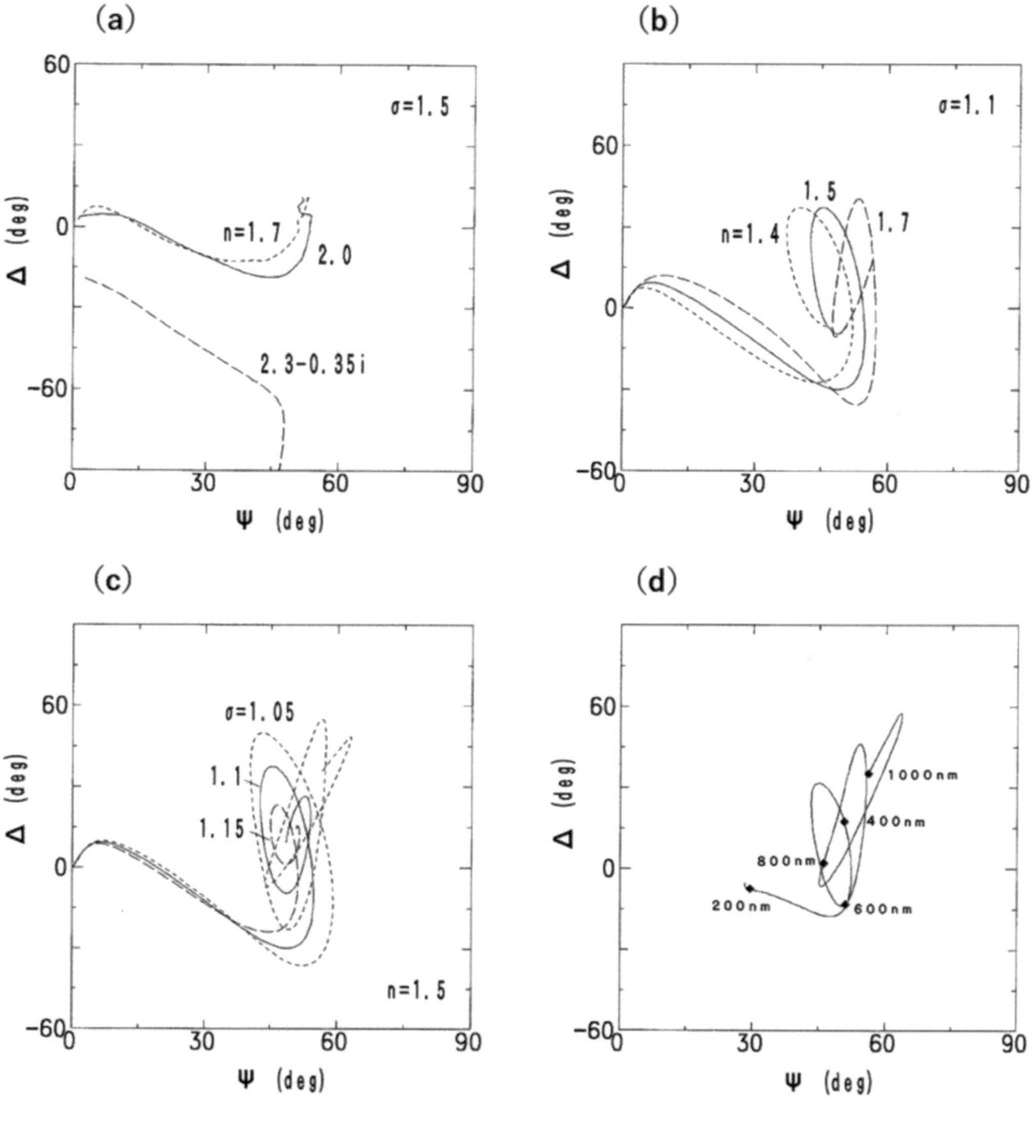


**Figure 8.** Evolution of the ellipsometric parameters by simulation: (a) the assumed geometric standard deviation of size distribution is 1.4, and the assumed refractive indexes are 2.3–0.35*i*, 2.0, and 1.7; (b) the assumed geometric standard deviation is 1.1, and the assumed refractive indexes are 1.7, 1.5, and 1.4; (c) the assumed geometric standard deviations are 1.15, 1.1, and 1.05, and the assumed refractive index is 1.5; (d) the translation-operated log-normal size distribution on a linear size scale is assumed with the refractive index of 1.5, and the geometric mean size and the geometric standard deviation are 50 nm and 2.0, respectively for the first lognormal distribution [2].

simulation for the growth of an ensemble of particles with a constant geometric standard deviation 1.5 for the lognormal size distribution, of which three different values of refractive index; 2.3–0.35*i*, 2.0, and 1.7, were assumed. The standard deviation and the first index are the values observed in the coagulation process of carbon particles as shown in 3.3.1. The last index is that of a hydrogenated amorphous carbon film at 488 nm [27]. As the particle diameter increases, the trajectory turned clockwise for absorbing particles, while counterclockwise for nonabsorbing particles. From the comparison of the experimental results with the calculation, it is seen that the growing particles are nonabsorbing and have the refractive index of a real number.

The calculated parameters converge to certain values while the experimental ones oscillate remarkably. Then, the calculation with a smaller geometric standard deviation of 1.1 was executed for refractive indexes of 1.7, 1.5, and 1.4 as shown in **Figure 8(b)**. It is seen that the trajectories draw loops. The inclination of the first loop of the experimental data is close to that with the index of 1.5.

**Figure 8(c)** shows the ellipsometric parameters in further evolution calculated in the case of a fixed refractive index of 1.5 with different geometric standard deviations of 1.15, 1.1, and 1.05. Comparing the results with the experimental trajectory, it can be seen that the standard deviation decreases with the increase of particle size, that is, the size distribution became a rather monodisperse one.

When fine particles grow by coating, size distribution function changes to that of translation-operated lognormal on a linear size scale as

$$N(D) = \frac{1}{\sqrt{2\pi}(D - D_c)\ln\sigma} e^{-\frac{\{\ln(D-D_c)-\ln D_m\}^2}{2(\ln\sigma)^2}}, \tag{8}$$

where $D_c$ means the thickness of coated material on the seeds of ultra-fine particles of the lognormal size distribution function as Eq. (6). **Figure 8(d)** shows the result of the simulation: the geometric mean size and the geometric standard deviation for the distribution of the injected particles are 50 nm and 2.0, respectively, and the refractive index is 1.5. The simulated trajectory agrees with the experimental results better than those in **Figure** 8(a–c). The deviation of the trajectory from the experimental one at the start may be due to the difference of the refractive index between hydrogenated carbon and pure carbon.

**Figure 9(a)** shows the variation of the value of $\Delta$ with time for the coating growth of carbon under conditions similar to those shown above. The best-fitted trajectory by simulation was obtained for the geometric mean size and the geometric standard deviation for the distribution of the injected particles is 50 nm and 1.5, respectively, and the refractive index is 1.53. The variation of $\Delta$ value with time for the best-fitted trajectory is shown in **Figure 9(b)** [3]. Characteristics of the two curves in the peaks and dips correspond well to each other. The growing particle diameter was determined from the correspondence. The time evolution of the particle diameter so determined is plotted in **Figure 10**. It is seen that the diameter increases almost linearly for about the first 1500 seconds, then the growth rate decreases gradually. This suggests that the state of particle plasma has changed at about 1500 seconds The density of

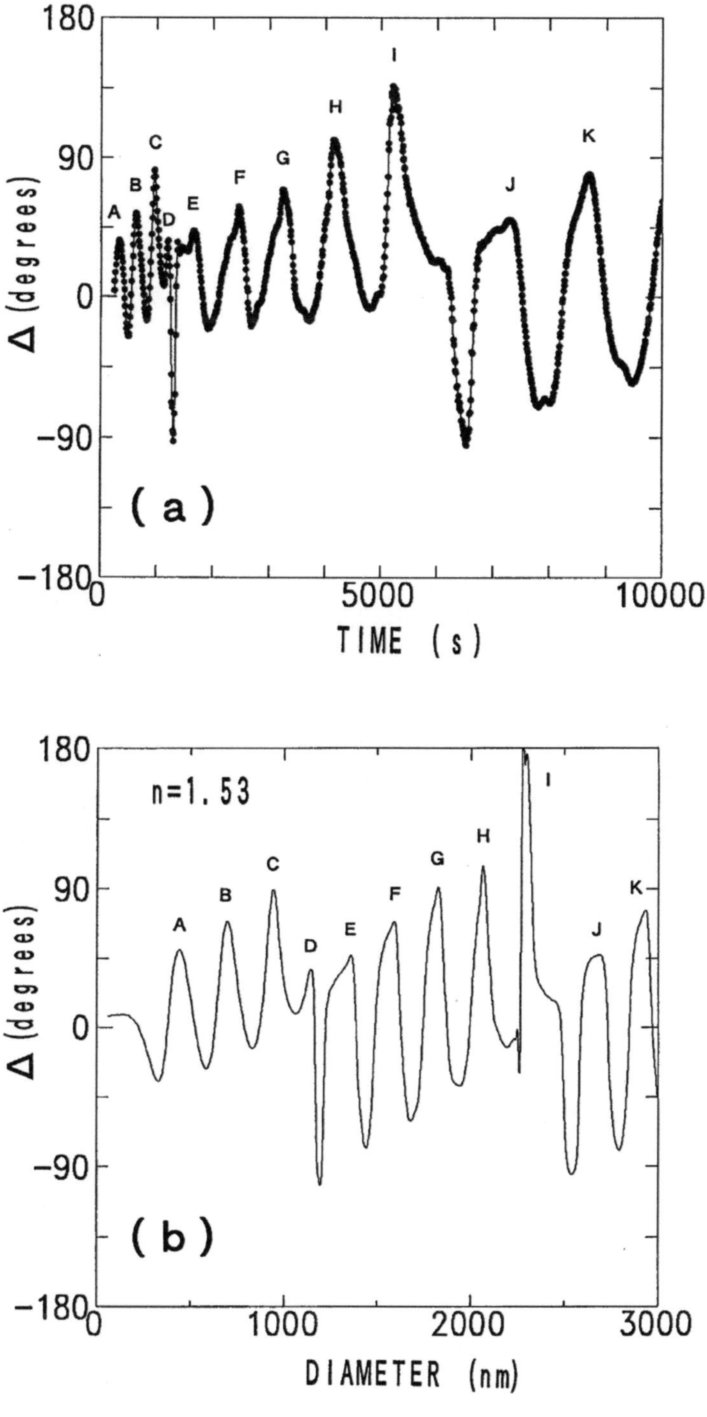


**Figure 9.** Experimentally obtained evolution of the ellipsometric parameter $\Delta$ with time (a), and calculated $\Delta$ as a function of the particle diameter with the refractive index 1.53 (b) [3]. The correspondence of the peaks is indicated by the same letters in both figures.

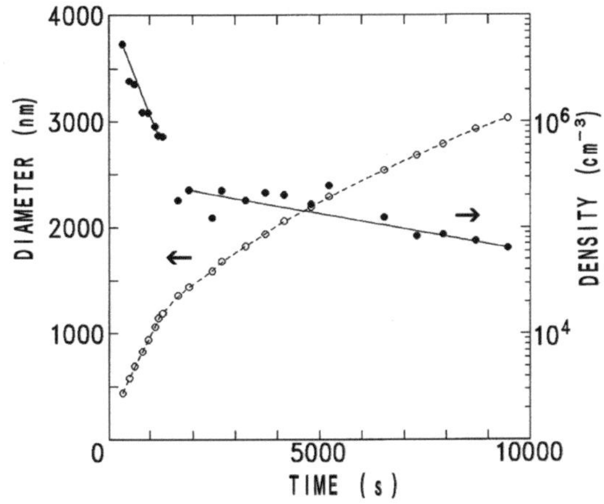


**Figure 10.** Evolution of particle diameter (open circles and a broken curve) and density (closed circles and solid lines) with time [3].

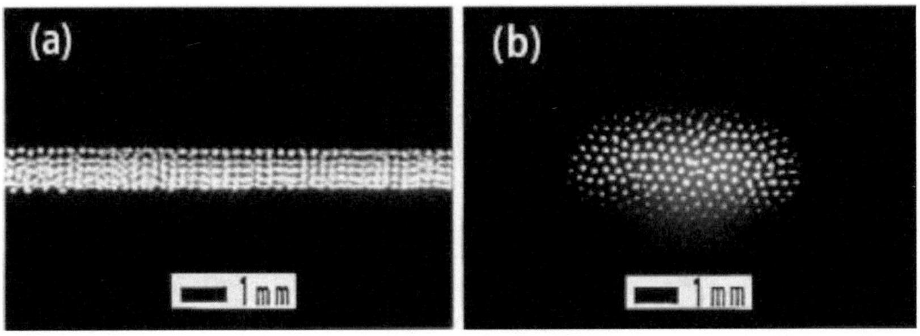


**Figure 11.** Photographs of a Coulomb crystal viewed from the side port (a) and the upper port (b) of the reactor [3].

particles is determined from the scattered light intensity, which is calibrated by the density evaluated from the interparticle distances in the photo of the Coulomb crystal formed in the plasma as shown in **Figure 11** [3]. It is seen in **Figure 10** that the density decrease is rapid for the first 1500 seconds, but subsequently becomes slow. It is speculated that the particles escape laterally from the plasma region through the side of the electrode gap before they are bound to one another, but they cannot escape easily after the Coulomb crystal is formed. From the consideration, it can be said that the phase transition from the state of "liquid" to that of "solid" [26] occurs at around 1500 seconds after the particle injection. At the point, the particle diameter is 1300 nm and the density is $3 \times 10^5$ cm$^{-3}$. The transition was also confirmed from the magnified photographs taken at times later than 1800 s, in which the particles were observed to be gradually arranged. The Wigner-Seitz radius is calculated from the density $N$ as $[3/(4\,\pi N)]^{1/3}$ [28], and becomes about 90 μm at the phase transition.

The particles suspended in plasma were observed after the growth using a scanning electron microscope (SEM). All particles were single spheres. **Figure 12** shows a SEM micrograph for the particles suspended in plasma for 3 h. It is seen that the particles are nearly spherical and equal in size with the diameter of 3.0–3.1 μm. The spherical shape supports the former discussion on Mie-scattering ellipsometry and coating growth mechanism.

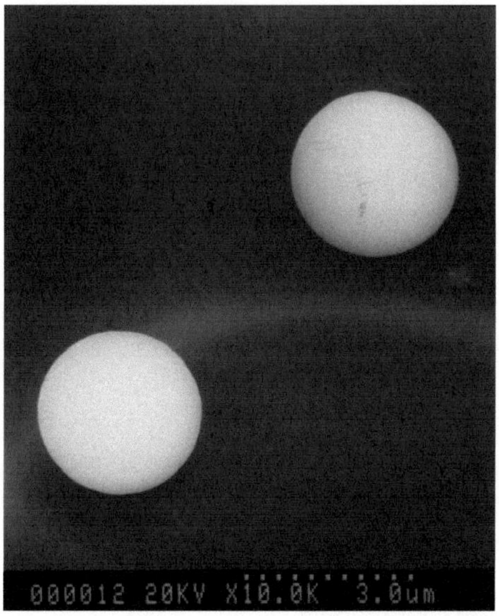


**Figure 12.** Typical SEM micrograph of particles suspended in plasma for 3 h.

From these results, it can be speculated about the growth of the carbon particles as follows. Hydrogenated amorphous carbon coated the ultra-fine carbon particles in methane plasma. About the smaller value of refractive index than that of a hydrogenated amorphous carbon, it can be understood that the void of 25% volume fraction was included in the coated carbon. Without the homogeneous nucleation of carbon particles in the plasma, which was confirmed by the fact that the Mie-scattered light is not observed in methane plasma without the injection of carbon ultra-fine particles under the same plasma conditions, the particles were equally coated independent of their initial size. From the spherical shape of the grown particles, it is imagined that the particles were coated isotropically. These grown particles with diameter of more than 3 μm can be suspended in methane plasma.

# 4. Imaging Mie-scattering ellipsometry

Imaging Mie-scattering ellipsometry, like the imaging ellipsometry for thin film analysis, enables the analysis of spatial distribution of particle size, as well as the easier confirmation of optical adjustment from a long distance. The difference between imaging Mie-scattering ellipsometer and above-mentioned Mie-scattering ellipsometer is the imaging of the distribution of fine particles using a 2D image sensor instead of a photodetector.

## 4.1. Experimental setup for imaging Mie-scattering ellipsometry

Ellipsometric measurement by a system with a rotating compensator has an advantage of the determination of all four Stokes parameters in one rotation. The light intensity changes by the rotation of compensator, whose azimuth is $C$, is expressed by sinusoidal function as [29, 30],

$$\mathscr{I}(C) = A_0 + A_2 \cos 2C + B_2 \sin 2C + A_4 \cos 4C + B_4 \sin 4C. \tag{9}$$

When the azimuth angle of the polarizer (P) and that of the analyzer (A) are 90 and ±45°, respectively, and the rotating compensator is put between the polarizer and particles, Fourier coefficients indicated in Eq. (9) are related with $\Psi$ and $\Delta$ for particles as,

$$\Psi = \frac{1}{2}\tan^{-1}\frac{\sqrt{{B_4}^2 + \frac{1}{4}{B_2}^2}}{A_4} \quad \text{for } A_4 \geqq 0, \text{ otherwise} \oplus 90°, \tag{10a}$$

$$\Delta = \tan^{-1}\frac{B_2}{2B_4} \quad \text{for } B_4 < 0(\mathrm{A} = 45°);\ B_4 \geqq 0(\mathrm{A} = -45°), \text{ otherwise} \oplus 180°. \tag{10b}$$

The system for the experimental example of imaging Mie-scattering ellipsometry is shown in **Figure 13** [12]. The ellipsometer consists of polarizer and analyzer modules and a 2D image sensor. The polarizer module includes a 532 nm wavelength laser, a polarizer (P = 90°), and a rotating compensator. The analyzer module includes a wire grid polarizer (A = −45°) and a mirror. A digital video camera containing a 2D image sensor for the detection of 2D distributed

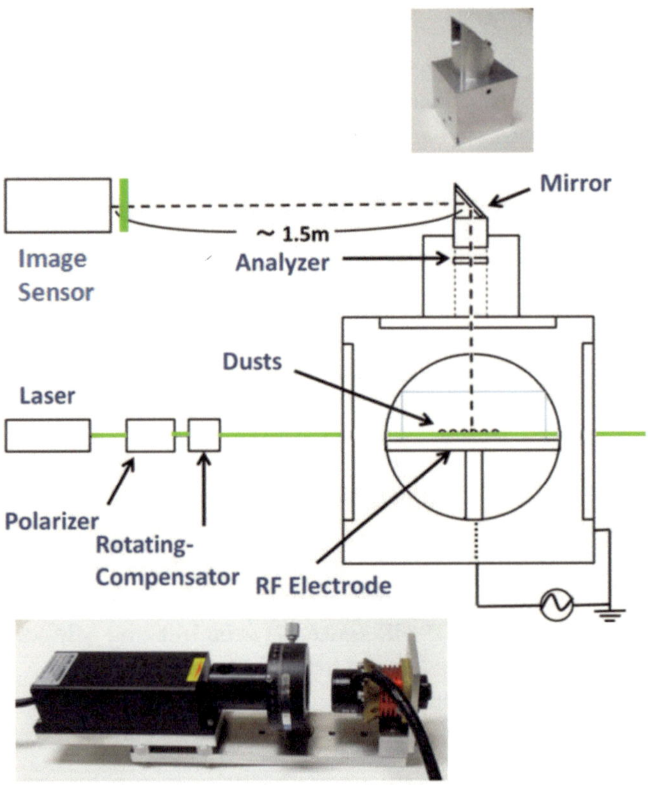


**Figure 13.** Schematic of ellipsometry measurement with dust plasma chamber. Inserted upper and lower photos are analyzer module and polarizer module, respectively [12].

light intensity was placed 1.5 m apart from the analyzer module. An optical band-pass filter with 1 nm FWHM at 532 nm is attached in front of the video camera. The images of distribution of fine particles by scattered light are expanded and focused using built-in functions and an extra lens. Thirty 2D images were captured every second. Although a rotating compensator is set in analyzer modules in conventional rotating-compensator ellipsometers [31, 32], it was set in the polarizer module in this case for the analyzer module able to be installed in a vacuum vessel.

The polarizer module and the digital video camera can be attached to view windows of a large plasma device in the atmosphere. The analyzer module is prepared to be fixed in a vacuum vessel [12]. It was ascertained that an image reflected by the mirror in the analyzed module could be viewed and recorded with the video camera out of the vacuum vessel of the large helical devise (LHD), a nuclear fusion experimental devise in Toki in Japan, when laser light was scattered by a small test material placed at the point of dust particle measurement.

### 4.2. Experiments of imaging Mie-scattering ellipsometry

In order to confirm the functional capability of the ellipsometric system, a preliminary experiment was carried out using a small dust plasma cubic chamber, 15 cm on a side [33]. Spherical divinylbenzene polymer particles of 2.27 μm in diameter with a dispersion degree of 0.1 μm were injected and suspended in argon plasma under a pressure of 50 Pa.

Light intensity data for 1600 pixels in the area covering the light scattering image were less than 75 of the maximum value of 255. Average values, i.e., total summed values of the data divided by pixel number, were calculated for images every 6° of compensator azimuth angle. The average light intensity ($C_n$) was obtained for each provisional azimuth angle $C_n$, where $C_n = 6n$. Then, Fourier analysis was performed and coefficients were obtained. After the difference between true compensator azimuth angle $C$ and provisional one $C_n$ was calculated through the obtained Fourier coefficients, true Fourier coefficients were calculated as

$$A_0 = 6.9009, \quad A_2 = 0, \quad B_2 = -0.9281, \quad A_4 = -1.2814, \quad B_4 = -0.6747. \tag{11}$$

Finally, ellipsometric parameters are determined by Eqs. (6a) and (6b) as

$$\Psi = 73.7^\circ, \quad \Delta = 214.5^\circ. \tag{12}$$

These values are plotted in $\Psi$-$\Delta$ coordinate (**Figure 14**). Meanwhile, $\Psi$ and $\Delta$ was calculated based on the Mie scattering theory for the values of the refractive index of 1.56 (for divinylbenzene), the diameter of 2.22–2.30 μm, and the scattering angle of 90°. The calculated values are also shown in **Figure 14**. It can be stated that they agree fairly well with each other and that the measurement method is reliable. This method has the advantage of easy optical adjustment from a long distance observing the image of fine particles.

Using the same system of imaging Mie-scattering ellipsometry, the spatial size distribution of parallelly layered fine particles in argon plasma was analyzed [34]. **Figure 15** shows a video image of light scattered by spherical divinylbenzene particles of 2.74±0.09 μm in diameter

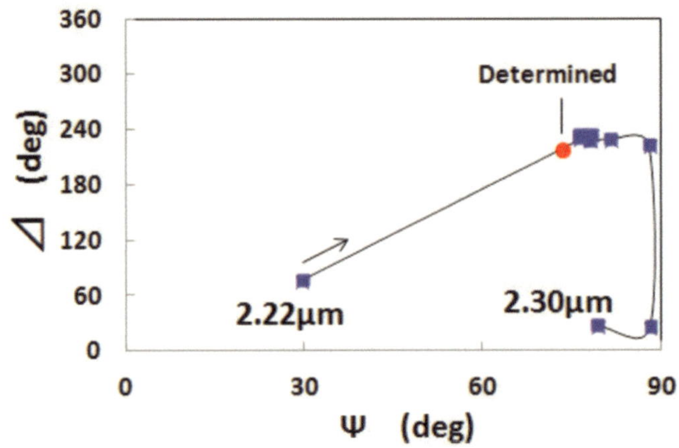


**Figure 14.** Ellipsometric parameters $\Psi$ and $\Delta$ determined by measurement (circle) and calculation (squares). Calculation was carried out based on Eq. (1) and the Mie-scattering theory for the values of refractive index of 1.56, the diameter of 2.22–2.30 μm by 0.01 μm and the scattering angle of 90° [12].

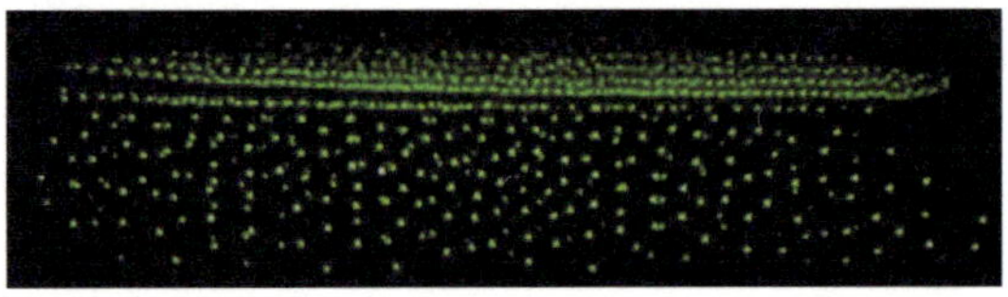

**Figure 15.** Video image of light scattered by spherical divinylbenzene particles of 2.74 ± 0.09 μm in diameter suspended in argon plasma at 65 Pa.

suspended in argon plasma with the pressure of 65 Pa. It is seen that the particles form vertical strings in the lower region and horizontal layers parallelly in the upper region. Fine particles more than 10 in each parallel layers were analyzed by imaging Mie-scattering ellipsometry for each area of 1320 pixels in video images. By the comparison of the determined values $\Psi$ and $\Delta$ with calculation, the size was evaluated to be 2.70, 2.74, 2.75, and 2.77 μm for particles in upper to lower layers.

## 5. Conclusion

The principium of Mie-scattering ellipsometry and its analytical method were explained in detail. Then, the results of analyses by in-process Mie-scattering ellipsometry for carbon particle growth in argon plasma, as well as in methane plasma, were shown.

In argon plasma, it was suggested from the simulated trajectory best fit to the experimental data that the carbon particles included voids and the size distribution was that of polydisperse particles, such as the diameter ranging mostly from 70 to 150 nm for the mean size of 100 nm. The mean diameter and the density of particles were evaluated, and their time dependence suggests that the size increase of the particles is mostly due to coagulation.

In the case of methane plasma, the trajectory obtained by the model of translation-operated lognormal size distribution on a linear scale agreed with experimental result better than that of lognormal of a constant geometric standard deviation. It was found that ultra-fine carbon particles were coated by hydrogenated amorphous carbon during suspension in methane plasma. The particles were equally coated independent of their initial size without homogeneous nucleation in the plasma. The spherical shape of carbon particles observed in SEM image supports that the particles were coated isotropically with carbon.

An imaging Mie-scattering ellipsometry system, consisting a laser, a polarizer, a rotating compensator, an analyzer and a 2D image sensor, was developed. It was shown that the use of a 2D image sensor instead of a photodetector enables easy confirmation of optical adjustment from a long distance by the image of fine particles without any problems for measurement. The size of distributed fine particles was determined for each separated area. The developed system of imaging Mie-scattering ellipsometer has the potential for the easier confirmation of optical adjustment from a long distance, as well as for the analysis of spatial distribution of particle size.

## Acknowledgements

The authors thank Mr. Masahiro Kawano and Mr. Toshihiko Arayama for experimental assistance, and Prof. Suguru Masuzaki and Prof. Kunihide Tachibana for useful discussions. This work was partly supported by the Ministry of Education, Science, Sports and Culture of Japan under Grant Number 24244094.

## Author details

Yasuaki Hayashi* and Akio Sanpei

*Address all correspondence to: hayashiy@kit.ac.jp

Kyoto Institute of Technology, Kyoto, Japan

## References

[1] Hayashi Y, Tachibana K. Mie-scattering ellipsometry for analysis of particle behaviors in processing plasmas. Japanese Journal of Applied Physics. 1994;**33**(3B):L476-L478. DOI: 10.1143/JJAP.33.L476

[2] Hayashi Y, Tachibana K. Analysis of spherical carbon particle growth in methane plasma by Mie-scattering ellipsometry. Japanese Journal of Applied Physics. 1994;**33**(7B):4208-4211. DOI: 10.1143/JJAP.33.4208

[3] Hayashi Y, Tachibana K. Observation of Coulomb-Crystal formation from growing particles grown in a methane plasma. Japanese Journal of Applied Physics. 1994;**33**(6A):L804-L806. DOI: 10.1143/JJAP.33.L804

[4] Chu JH, Lin I. Direct observation of Coulomb crystals and liquids in strongly coupled rf dusty plasmas. Physical Review Letters. 1994;**72**(25):4009-4012. DOI: 10.1103/PhysRevLett.72.4009

[5] Thomas H, Morfill GE, Demmel V, Goree J, Feuerbach B, Möhlmann D. Plasma crystal: Coulomb crystallization in a dusty plasma. Physical Review Letters. 1994;**73**(5):652-655. DOI: 10.1103/PhysRevLett.73.652

[6] Tachibana K, Hayashi Y. Analysis of the Coulomb-solidification process in particle plasma. Australian Journal of Physics. 1995;**48**:469-477. DOI: 10.1071/PH950469

[7] Hayashi Y, Tachibana K. Coulomb crystal formation from growing particles in a plasma and the analysis. Journal of Vacuum Science and Technology. 1996;**14**(2):506-510. DOI: 10.1116/1.580116

[8] Hong SH, Winter J. Size dependence of optical properties and internal structure of plasma grown carbonaceous nanoparticles studies by in situ Rayleigh-Mie scattering ellipsometry. Journal of Applied Physics. 2006;**100**:064303. DOI: 10.1063/1.2338132

[9] Groth S, Greiner F, Tadsen B, Piel A. Kinetic Mie ellipsometry to determine the time-resolved particle growth in nanodusty plasmas. Journal of Physics D: Applied Physics. 2015;**48**:465203. DOI: 10.1088/0022-3727/48/46/465203

[10] Shiratani M, Kawasaki H, Fukuzawa T, Yoshioka T, Ueda Y, Singh S, Watanabe Y. Simultaneous in situ measurements of properties of particulates in rfsilane plasmas using a polarization-sensitive laser-light-scattering method. Journal of Applied Physics. 1996;**79**(1):104-109. DOI: 10.1063/1.360916

[11] Greiner F, Carstensen J, Kohler N, Pilch I, Ketelsen H, Knist S, Piel A. Imaging Mie-ellipsometry: Dynamics of nanodust clouds in an argon-acetylene plasma. Plasma Sources, Science and Technology. 2012;**21**(6):065005. DOI: 10.1088/0963-0252/21/6/065005

[12] Hayashi Y, Kawano M, Sanpei A, Masuzaki S. Mie-scattering ellipsometry system for analysis of dust formation process in large plasma device. IEEE Transactions on Plasma Science. 2016;**44**(6):1032-1035. DOI: 10.1109/TPS.2016.2542349

[13] Kerker M. The use of white light in determining particle radius by the polarization ratio of the scattered light. Journal of Colloidal Science. 1950;**5**:165-167

[14] Azzam RMA, Bashara NM. Ellipsometry and Polarized Light. Amsterdam: North Holland; 1987. p. 539

[15] Theeten JB, Hottier F, Hallais J. Ellipsometric assessment of (Ga, Al) As/GaAs epitaxial layers during their growth in an organometallic VPE system. Journal of Crystal Growth. 1979;**46**(2):245-252. DOI: 10.1016/0022-0248(79)90064-2

[16] Theeten JB. Real-time and spectroscopic ellipsometry of film growth: Application to multilayer systems in plasma and CVD processing of semiconductors. Surface Science. 1980;**96**(1-3):275-293. DOI: 10.1016/0039-6028(80)90307-6

[17] Hayashi Y, Itoh A. Ellipsometric monitor for film thickness control in plasma processes. In: Proceedings of International Ion Engineering Congress, ISIAT'83 & IPAT'83; Kyoto, Japan. 1983. pp. 1469-1474

[18] Collins RW, Pawlowski A. The nucleation and growth of glow-discharge hydrogenated amorphous silicon. Journal of Applied Physics. 1986;**59**(4):1160-1166. DOI: 10.1063/1.336553

[19] Antoine AM, Drevillon B, Roca i Cabarrocas P. In situ investigation of the growth of rf glow-discharge deposited amorphous germanium and silicon films. Journal of Applied Physics. 1987;**61**(7):2501-2508. DOI: 10.1063/1.337924

[20] Hayashi Y, Drawl W, Collins RW, Messier R. In-process ellipsometric monitoring of diamond film growth by microwave plasma enhanced chemical vapor deposition. Applied Physics Letters. 1992;**60**(23):2868-2870. DOI: 10.1063/1.106827

[21] Hayashi Y, Drawl W, Messier R. Temperature dependence of nucleation density of chemical vapor deposition diamond. Japanese Journal of Applied Physics. 1992;**31**(2B):L193-L196. DOI: 10.1143/JJAP.31.L193

[22] van de Hulst HC. Light Scattering by Small Particles. New York: Dover; 1981. p. 470

[23] Tachibana K, Hayashi Y, Okuno T, Tatsuta T. Spectroscopic and probe measurements of structures in a parallel-plates RF discharge with particles. Plasma Sources Science and Technology. 1994;**3**(3):314-319. DOI: 10.1088/0963-0252/3/3/012

[24] Deirmendjian D. Electromagnetic Scattering on Spherical Polydispersions. New York: American Elsevier; 1969. p. 290

[25] Yoshida T, Okuyama K, Kousaka Y, Kida Y. Change in particle size distributions of polydisperse aerosols undergoing Brownian coagulation. Journal of Chemical Engineering of Japan. 1975;**8**(4):317-322. DOI: 10.1252/jcej.8.317

[26] Hayashi Y. Thin film monitoring with ellipsometry in in-line processing equipment. Japanese Journal of Applied Physics. 1990;**29**(11):2514-2518. DOI: 10.1143/JJAP.29.2514

[27] Ikezi H. Coulomb solid of small particles in plasmas. The Physics of Fluids. 1986;**29**:1764-1766. DOI: 10.1063/1.865653

[28] Ichimaru S. Strongly coupled plasmas: High-density classical plasmas and degenerate electron liquids. Reviews of Modern Physics. 1982;**54**(4):1017-1059. DOI: 10.1103/RevModPhys.54.1017

[29] Hauge PS, Dill FH. A rotating-compensator Fourier ellipsometer. Optics Communications. 1975;**14**(4):431-437. DOI: 10.1016/0030-4018(75)90012-7

[30] Hauge PS. Generalized rotating-compensator ellipsometry. Surface Science. 1976;**56**:148-160. DOI: 10.1016/0039-6028(76)90442-8

[31] Lee J, Rovira PI, An I, Collins RW. Rotating-compensator multichannel ellipsometry: Applications for real time Stokes vector spectroscopy of thin film growth. Review of Scientific Instruments. 1998;**69**(4):1800-1810. DOI: 10.1063/1.1148844

[32] Weiß R, Hong S-H, Ränsch J, Winter J. Rayleigh. Mie scattering ellipsometry as an in situ diagnostic for the production of 'smart nanoparticles'. Physica Status Solidi A. 2008;**205**(4): 802-805. DOI: 10.1002/pssa.200777803

[33] Hayashi Y, Mizobata Y, Takahashi K. Experiments of fine-particle plasma using planar magnetron plasma system. International Journal of Microgravity Science and Application. 2011;**28**(2):S23-S26

[34] Hayashi Y, Sanpei A, Mieno T, Masuzaki S. Analysis of spatial distribution of fine particles in plasma by imaging Mie-Scattering ellipsometry. In: 8th International Conference on the Physics of Dusty Plasmas; 20-25 May 2017; Prague, Czech Republic

Chapter 2

# Achromatic Ellipsometry: Theory and Applications

Eralci Moreira Therézio, Gustavo G. Dalkiranis,
André A. Vieira, Hugo Gallardo, Ivan H. Bechtold,
Patricia Targon Campana and Alexandre Marletta

Additional information is available at the end of the chapter

http://dx.doi.org/10.5772/intechopen.70089

**Abstract**

In the present chapter, the theory and some applications of Achromatic Ellipsometry, including transmittance, absorbance, and emission, are presented. The new methodology introduced here comprises the calculation of Stokes parameters using Fourier series analysis. Light polarization was determined by calculating the polarization degree, anisotropy, asymmetry parameters, and rotational and ellipsometry angles. The nematic liquid crystal E7™ doped with 4,7-bis{2-[4-(4-decylpiperazin-1-yl) phenyl]ethynyl}-[2,1,3]-benzothiadiazole (5A) within twisted and parallel structures, was used to illustrate the applications for this technique, that has been shown to be an innovative and versatile tool to correlate the photophysics with materials structure.

**Keywords:** ellipsometry, optical spectroscopy, Stokes parameters

## 1. Introduction

The understanding of light polarization from materials absorption, emission, and reflection can be addressed by matching one or more polarizers in the optical path of conventional photoluminescence and UV-vis spectrometers, called polarized photoluminescence and polarized absorbance, respectively [1–4]. These methodologies have been applied with the aim of acquiring information on the molecular arrangement (anisotropy and order factor) [3–6]. Besides that, Circular Dichroism spectroscopy has also been widely used to understand molecular organization based on their chirality [2, 6, 7]. Nevertheless, these techniques fail to identify all possible polarization states when applied separately. In the case of photoluminescence, for example, it is possible to get information from linear polarization states, but not from circular ones. The polarized absorption, on the other hand, can elucidate the molecular order, but does not provide information on circular birefringence effects and Circular Dichroism. The last one consists in the difference between left and right circularly polarized light, thus resulting in the molecular asymmetry [6, 8, 9]. For instance, the study of energy transfer processes by means of the investigation of emitted polarized light can be very complicated when the polarized state is not completely described.

In this sense, Emission Ellipsometry (EE) is a technique that, combined with the Stokes theory, allows the complete description of the emitted light polarization states [6, 8, 9] being, in this manner, an excellent alternative in the study of energy transfer processes, mainly in polymeric materials [2, 5, 10, 11]. In addition, it is possible to obtain full information regarding molecular organization [2, 12], anisotropy [10], and asymmetry [13] factors in photoluminescent materials. All these studies and several applications in diverse materials lead to the development of a new method for Raman Optical Activity studies, named ellipsometric Raman spectroscopy (ERS) [14, 15]. In this new methodology, Basilio et al. [14] have been shown that EE combined with the Stokes theory can also be applied, not only for emitted but also for the scattered light. From this experiment, one can obtain meaningful information about materials optical activity even for easily photodegradable ones (both biological and polymeric) due to the limited exposition to exciting light.

This chapter aims a detailed review of Achromatic Ellipsometry from the appliance of Stokes theory on the determination of polarization parameters to its applications on materials studies.

## 2. Stokes theory

In the middle of the nineteenth century, the light polarization has had been mathematically characterized by Augustin Jean Fresnel (1788–1827) and Dominique François Arago (1786–1853), among others. Notwithstanding, the concept of partially polarized light was only formulated until the period around the year 1852, by Sir George Gabriel Stokes [12]. Fresnel and Arago used to describe polarized light in terms of its polarization ellipse. Such representation was incomplete, because it did not include partially polarized light. Stokes, on the other hand, solved the problem by demonstrating that it was possible to completely characterize any arbitrary polarization state from four experimental measurements (observables), in his work "On the Composition and Resolution of Streams of Polarized Light from Different Sources" [16]. Here and now, these observables are designated as Stokes parameters, which describe nonpolarized, partially, and complete polarized lights. Despite its fundamental contribution, his work was practically ignored by the scientific community for almost 100 years. In 1946, the Nobel laureate Subrahmanyan Chandrasekhar formulated the equations for radioactive transfer of partially polarized light [12], which was only possible due to Stokes' work. In addition to Stokes theory, new material developments for achromatic optical elements made possible the measuring of all polarization states as a function of wavelengths (optical spectrum), i.e., the necessary observables to enlighten the materials structure [2, 13, 14, 17, 18].

### 2.1. Polarization ellipse

Light description in terms of polarization ellipse, which is valid as long as the studied light is completely polarized, is very convenient. It allows the characterization of several states of light polarization using one equation [14, 18] and can result in forms interpreted as linearly and circularly polarized light. Beyond that, from the concept of polarization ellipse, it is possible to understand Stokes' description to identify all states of light polarization, including partially polarized beams [12]. In this sense, polarized light propagating along the $z$ direction, and the electric field in two orthogonal components propagating along $x$ and $y$ directions can be represented by the following equations:

$$E_x(z,t) = E_{0x} \cos(\tau + \delta_x) \tag{1}$$

$$E_y(z,t) = E_{0y} \cos(\tau + \delta_y) \tag{2}$$

where $E_{0x}$ and $E_{0y}$ are the maximum amplitude, $\delta_x$ and $\delta_y$ are the phase constants at $x$ and $y$ axis respectively, and $\tau = \omega t - kz$ is the diffuser. From Eqs. (1) and (2), one can obtain the equation for an arbitrary instant of time, $t$:

$$\frac{E_x^2}{E_{0x}^2} + \frac{E_y^2}{E_{0y}^2} - 2\frac{E_x E_y}{E_{0x}E_{0y}} \cos\delta = \sin^2(\delta) \tag{3}$$

where $\delta = \delta_y - \delta_x$. Equation (3) is named polarization ellipse [12, 19]. The ellipse described by Eq. (3) has its $a$ and $b$ axis rotated by an angle $\Psi$ relatively to the reference plan $xy$, as can be seen in **Figure 1**.

From the diagram above (**Figure 1**), it is possible to write the components $E'_x$ and $E'_y$ relatively to the components $E_x$ and $E_y$ and to the angle $\Psi$, in a way that:

$$E'_x = a\cos(\tau + \delta') \tag{4}$$

$$E'_y = \pm b\sin(\tau + \delta') \tag{5}$$

$$\frac{E'^2_x}{a^2} + \frac{E'^2_y}{b^2} = 1 \tag{6}$$

Equation (6) describes a standard ellipse, straight, where $a$ and $b$ are the major and minor axis, respectively [2, 12, 19]. It is important to note that, when the phase angle equals 90° and 270°, the angle $\Psi$ will be zero. This condition is only reached when $E_{0x}$ and $E_{0y}$ are zero, that is, for linearly polarized light along vertical ($y$-axis) or horizontal ($x$-axis) directions. Another important parameter is defined as ellipticity ($\chi$) [5, 12]:

$$\tan(\chi) = \pm\frac{a}{b} \tag{7}$$

In this case, if $a = b$, which means that the light is left or right circularly polarized or even random, then the ellipticity became zero.

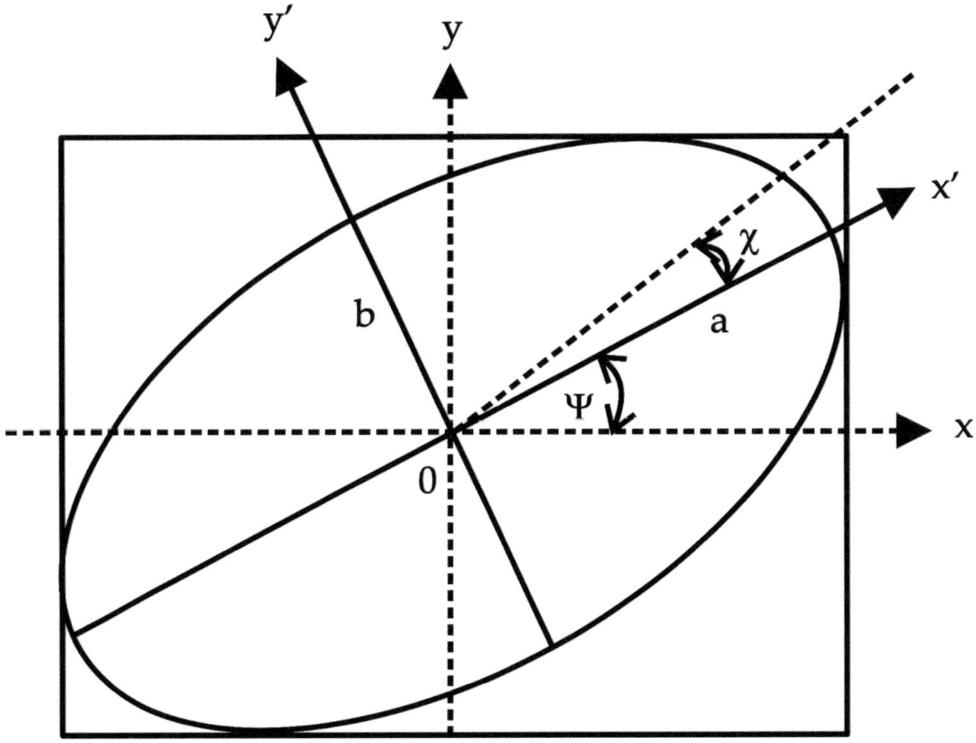


**Figure 1.** Polarization ellipse representation.

### 2.2. Stokes parameters

It is well known that the amplitude of the optical field cannot be observed [12, 15, 16, 19]. However, its intensity, derivative from the temporal average of square amplitude, can be observed. Therefore, one can obtain the polarization ellipse observables by taking the temporal average of nonobservables from the polarization ellipse. By doing that, it is possible to collect four parameters, the Stokes parameters [12, 16]. Consequently, Stokes parameters are a natural outcome of the wave theory of light, in a way that provide a complete description of any light polarization state [12]. More important, they are exactly the measured quantities.

Formerly, the Stokes parameters were used only for describing intensity measurements and polarization states of optical fields. Still, when these parameters are found in the form of column matrix (also known as Stokes vectors) they can lead to the study of spectroscopic observables [11–14, 17, 18]. Therein formalism, in Eq. (6), each term represents the temporal average, such as in Eq. (8) in its general form [19]:

$$\frac{E_x^2(t)}{E_{0x}^2} + \frac{E_y^2(t)}{E_{0y}^2} - 2\frac{E_x(t)E_y(t)}{E_{0x}E_{0y}}\cos\delta = \sin^2(\delta) \tag{8}$$

and the temporal average for each term is given by [12, 16]:

$$E_x^2(t) = \frac{1}{2}E_{0x}^2, \quad E_y^2(t) = \frac{1}{2}E_{0y}^2, \text{ and } E_x(t)E_y(t) = \frac{1}{2}E_{0x}E_{0y}\cos\delta \tag{9}$$

Combining the equations (8) and (9), one can extract the Stokes parameters[1] from the following equation:

$$(E_{0x}^2 + E_{0y}^2)^2 - (E_{0x}^2 - E_{0y}^2)^2 - (2E_{0x}E_{0y}\cos\delta)^2 = (2E_{0x}E_{0y}\sin\delta)^2 \tag{10}$$

arranged as follows:

$$\begin{aligned} S_0 &= E_{ox}^2 + E_{oy}^2 \\ S_1 &= E_{ox}^2 - E_{oy}^2 \\ S_2 &= 2E_{0x}E_{0y}\cos\delta \\ S_3 &= 2E_{0x}E_{0y}\sin\delta \end{aligned} \tag{11}$$

Using the parameters above, Eq. (10) can be rewritten in terms of complete polarized light:

$$S_0^2 = S_1^2 + S_2^2 + S_3^2 \tag{12}$$

The Stokes parameters, resulted from the definitions in Eq. (11), are real quantities expressed in terms of intensity. More than that, they are the observables from the polarization ellipse and, therefore, they represent the optical field.

The first parameter, $S_0$, represents the total light intensity. The following parameters describe the differences between: (i) horizontal or vertical linear polarization – $S_1$, (ii) linear polarization at $\pm 45°$ – $S_2$, and (iii) left or right circular polarization – $S_3$. For partially polarized light, using the Cauchy-Schwarz Inequality [12, 16], Eq. (12) can be rewritten as follows:

$$S_0^2 \geq S_1^2 + S_2^2 + S_3^2 \tag{13}$$

where equality applies for completely polarized light and inequality for nonpolarized or partially polarized light. Since the Stokes parameters directly reflect the observables, one can represent the rotation angle ($\Psi$) and the ellipticity ($\chi$) using Eqs. (11) – (13) as:

$$\tan(2\Psi) = \frac{S_2}{S_1} \tag{14}$$

$$\sin(2\chi) = \frac{S_3}{S_0} \tag{15}$$

Finally, the degree of polarization ($P$) can be described using the Stokes parameters for any polarization state [2, 12, 19]:

---

[1]Stokes parameters can also be obtained by means of plane waves. This solution can be found at Refs. [12, 16, 19].

$$P = \frac{I_{pol}}{I_{tot}} = \frac{(S_1^2 + S_2^2 + S_3^2)^{\frac{1}{2}}}{S_0} \tag{16}$$

where $I_{pol}$ means intensity of the sum of polarized components and $I_{tot}$ is the total intensity of the light beam. The value $P = 1$ corresponds to completely polarized light, whereas $P = 0$ corresponds to nonpolarized and $0 < P < 1$ to partially polarized light.

### 2.3. How to measure Stokes parameters

As described in earlier sections, the Stokes parameters are real observables. To measure them, it is necessary at least two optical components with which the light must interact to be analyzed successfully. Generally, these elements are polarizers and quarter-wave plates that can be set up in several ways in order to acquire the Stokes parameters. These optical components can be mathematically represented as a Muller matrix [2, 12], and, from the mathematic relations, the Stokes parameters can be obtained (see Eq. (18)).

There are several available methods to get the Stokes parameters. Collet describes diverse optical configurations and measurement modes to calculate them [12]. Although, one must be careful with systematic errors, alignment and intensity corrections that can be introduced in the system depending on the chosen experimental assembly, mainly on those which the exchange of optical elements during the experiment is required. One way to avoid this problem is to keep the optical elements fixed, reducing a significant amount of errors from alignment and problems from light intensity corrections due to reflections or absorptions. This is the main proposal of this chapter: keeping the optical elements fixed, the observables will be decomposed by Fourier series. This method also provides the advantage of using achromatic optical elements that work in a broad spectral range, especially at visible spectrum.

The representation of light polarization states as Stokes parameters in the form of column matrix (Stokes vector) can be seen below:

$$S = \begin{pmatrix} S_0 \\ S_1 \\ S_2 \\ S_3 \end{pmatrix} \tag{17}$$

In this matrix formalism, the effects on the light polarization from any beam due to optical components can be mathematically represented also by matrix, in this case, the Muller matrix [12, 14]. Concerning the fixed optical components, the analysis method via Fourier series employs a quarter-wave plate and a fixed linear polarizer, as can be seen in **Figure 2**.

From the arrangement shown in **Figure 2**, it is possible to correlate the intensity of emerging field ($E'_x$ and $E'_y$) with the Stokes parameters from incident field ($E_x$ and $E_y$) through Eqs. (5), (14), and (16):

$$I(\theta) = \frac{1}{2}[A + B.\sin(2\theta) + C.\cos(4\theta) + D.\sin(4\theta)] \tag{18}$$

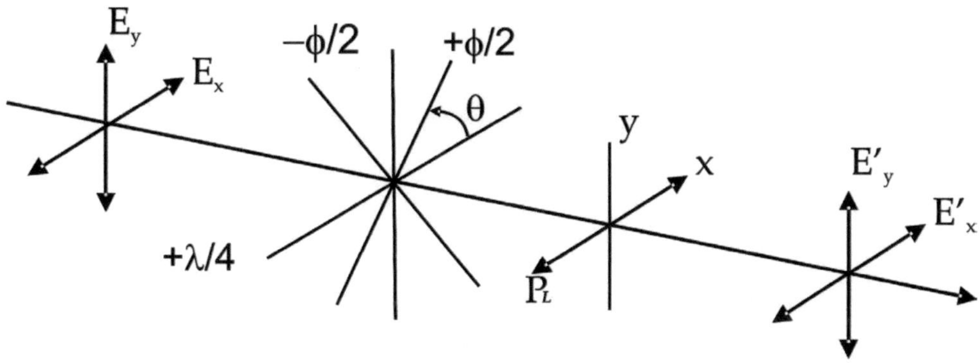


**Figure 2.** Diagram of an electromagnetic wave propagating through a quarter-wave plate, rotated by an angle θ, and a linear polarizer $P_L$ (horizontal) for the measurement of Stokes Parameters.

where for a polarizer with transmission axis along the horizontal direction ($x$-axis): $A = S_0 + \frac{S_1}{2}$, $B = -S_3$, $C = \frac{S_1}{2}$, and $D = \frac{S_2}{2}$; or a polarizer with transmission axis along the vertical direction ($y$-axis): $A = S_0 - \frac{S_1}{2}$, $B = S_3$, $C = \frac{-S_1}{2}$, and $D = \frac{-S_2}{2}$.

Eq. (18) is a truncated Fourier series, with a constant term $A$, a term with double frequency ($B$), and two more terms with quadruple frequencies ($C$ and $D$). As a Fourier series, this equation admits only one solution for parameters $A$, $B$, $C$, and $D$; thus resulting in a very precise way to obtain Stokes parameters. It is also likely to obtain the polarization degree ($P$) for the emerging light, from Eq. (18). In addition, Alliprandini-Filho [2], describes the asymmetry factor ($g$) and Circular Dichroism acquired from Stokes parameters, providing more complex information about the chirality of the molecules that compose the studied material. One can see that the technique described here is a powerful tool in the understanding of material conformations and structure.

### 2.3.1. *Anisotropy*

Similarly to the new approach as presented at previous sections, Therézio et al. [10] outlined a new method to determine anisotropy using Stokes parameters. In this methodology, the Stokes parameters are firstly rewritten as a function of electric field intensities, as described in Eq. (18). Then, the $y$ component is considered in the same direction as the laboratory frame of reference, and $x$ in the horizontal direction at the same reference. Besides, $E_{ox}^2 = I_{0x}$ and $E_{oy}^2 = I_{0y}$, where $I_{0i}$ is the beam intensity at $i$ direction. Then, considering that a sample is excited by a beam at $y$ direction is possible to obtain $E_{ox}^2 = I_{\parallel,\perp}$ and $E_{oy}^2 = I_{\parallel,\parallel}$, where $I_{\parallel,\perp}$ refers to the emission at perpendicular direction (⊥) due to parallel excitation (||). Hence, the first and second equations from Eq. (11) set can be expressed as:

$$\begin{cases} S_0 = I_{\parallel,\perp} + I_{\parallel,\parallel} \\ S_1 = I_{\parallel,\perp} - I_{\parallel,\parallel} \end{cases} \tag{19}$$

From the conventional equation that determines the anisotropy factor, $r$ [1, 10]:

$$r = \frac{I_{\parallel,\parallel} - I_{\parallel,\perp}}{I_{\parallel,\parallel} + 2I_{\parallel,\perp}}. \tag{20}$$

combined with Eq. (11), the anisotropy factor as a function of Stokes parameters is given by:

$$r = \frac{-2\frac{S_1}{S_0}}{3 + \frac{S_1}{S_0}}. \tag{21}$$

Perhaps the most advantage of Eq. (21), besides the obvious application as efficient technique for the studies of photophysical properties of luminescent materials, is the useless of corrections related to optical components, such as diffraction grade and slits when compared with the traditional methods as the one described by Eq. (20).

### 2.3.2. *Anisotropy and Circular Dichroism*

Despite the fact that Therézio et al. [13] had originally described this methodology for the study and characterization of a cholesteric liquid crystal, the dissymmetry and Circular Dichroism can be measured and studied with this method for any material of interest, since the material presents the measurable physical characteristics, as symmetry, for example.

Here the asymmetry factor, $g$, is defined by:

$$g = -2\frac{I^R - I^L}{I^R + I^L} \tag{22}$$

where $I^R$ and $I^L$ are the light intensities with excitation from right (R) and left (L) circularly polarized light. It is also related to parameter $\Delta$, the difference between the two circularized lights, and is expressed by:

$$\Delta = \frac{-g}{2} \tag{23}$$

The asymmetry factor becomes a Circular Dichroism measurement when the absorption of linearly polarized light across a sample is analyzed.

In the formulation presented here, $g$ can be determined from Stokes parameters. It can be noticed, in Eq. (19), that the Stokes parameter $S_3$ represents the difference $I^R - I^L$ and $S_0$ the sum of intensities $I^R + I^L$. As the linearly polarized light can be decomposed into the sum of two circularly polarized light beams (left and right), the Stokes parameters $S_3$ and $S_0$ can be directly related to $g$ parameter [13] by the following equation:

$$g = 2\frac{S_3}{S_0} \tag{24}$$

### 2.3.3. *Ellipsometric Raman Spectroscopy*

The Raman Optical Activity (ROA) has been provided numerous information about chiral molecules conformation. Those molecules have fundamental relevance in many areas of knowledge including Biology and Medicine. Basilio et al. [14] present details of a new technique to measure the Raman of chiral carbon: the ellipsometric Raman spectroscopy (ERS). The technique ERS quantifies the vibrational optical activity through measurement of small differences in intensity of Raman scattering for chiral molecules, when the impinging light is right and left circularly polarized. In other words, ERS is the application of Achromatic Ellipsometry to measure the Raman scattering. That is for say, through the technique of Ellipsometry one can measure the Stokes parameters, thus describing the complete light polarization states. The exciting light in ERS is a laser source which can be decomposed into two circular polarizations: right and left. Similarly to ROA experiments, the parameter $S_3$ describes the amount of light when normalized. ESR technique was validated by the successful characterization of the chiral alcohol (*S*)-(−)-1-phenylethanol (1-PhEtOH).

## 3. Applications

In this section, applications of Achromatic Ellipsometry will be briefly presented: the utilization of Emission Ellipsometry for studying a luminescent dye into a nematic liquid crystal. In this example, the importance of using the right optical elements will be noticed.

### 3.1. Nematic Liquid Crystal E7™ doped with a benzothiadiazole derivative dye

Samples were prepared using nematic liquid crystal E7™. Basically, it consists of a mixture of LC 5CB (47%), 7CB (25%), 8OCB (18%), and 5CT (10%), doped with a luminescent dye 4,7-bis {2-[4-(4-decylpiperazin-1-yl) phenyl]ethynyl}-[2,1,3]-benzothiadiazole (5A) at 0.025% of molar concentration [20]. Cells were prepared by superimposing two BK7 glass substrates coated with Poly(vinyl alcohol) (PVA) films separated by 20 mm thick Mylar spacers. The PVA films were previously rubbed with soft velvet to induce a specific aligning direction to the liquid crystal molecules. In this work, three different configurations were investigated considering the aligning direction induced in each glass plate of the cell: S00, S45 and S90 represent the rotation angles between the two substrates of the cells at 0° (parallel configuration), 45° and 90° (twisted), respectively. The alignment direction 0° was determined to be parallel with the vertical direction of lab reference. The scheme of the cells is presented in **Figure 3**.

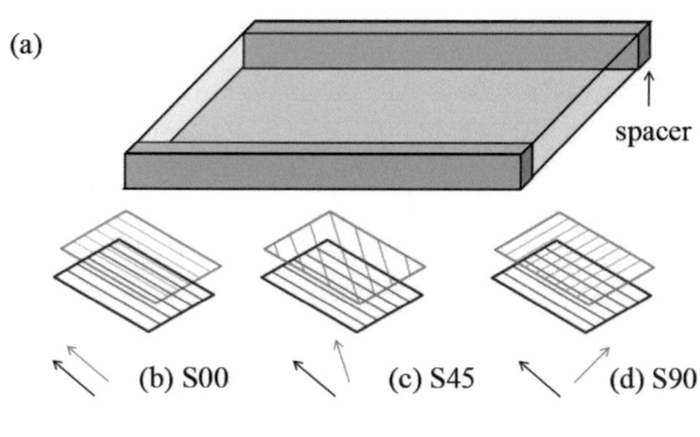


**Figure 3.** (a) Scheme of the liquid crystal cell. Rows indicate the alignment direction of each substrate of the cells (b) S00 – parallel axis (0°), (c) S45 – axis rotated in 45° and (d) S90 – perpendicular axis (90°).

Optical absorption (OA) measurements in UV-Vis-NIR were made in a FEMT O 800XI spectrometer. In the measurements of polarized optical absorption (POA), calcite polarizer Mugta10™ was inserted in the optical pathway of incident light, generating linear horizontally polarized light (LHP) and vertically polarized light (LVP). Photoluminescence measurements were made with an incident light of argon laser from Spectra Physics Inc. Stabilite 2017, with excitation wavelength at 457 nm. The emitted light from the sample was collimated by a set of lenses and detected by the Ocean Optics USB 2000 spectrometer. PL measurements were made on the two faces of the samples and the light was recorded on the same face.

Samples also followed EE and TE measurements. For the TE measurements, polarized light was analyzed by the Fourier series decomposition. The incident light was polarized vertically, right and left circularly. In the experimental set up, in the optical way of the transmitted light, an achromatic quarter-wave retarder was used, and a fixed calcite polarizer was placed before the spectrometer. In this case, a power meter of the OPHIR laser measurement group was used as a detector. The argon laser from Spectra Physics Inc. Stabilite 2017 at 457 nm and the He-Ne laser at 633 nm were used as light source. In this experiment, the quarter-wave retarder rotates around the propagation axis of transmitted light from 0 to 360° in steps of 10°.

**Figure 4a** presents absorbance and PL spectra normalized at maximum intensity of S00, S45 and S90 samples. Optical properties of the 5A dye did not change the electronics-vibrational transitions $\pi \rightarrow \pi^*$ e $\pi^* \rightarrow \pi$ significantly due to guest-host interaction in different cell configurations. The PL measurements were also performed with the wavelength excitation at 488 and 514 nm, and the emission did not show changes in the line shape or in the band position. In **Figure 4b–d**, polarized absorption spectrum, parallel, and perpendicular polarization are presented for S00, S45, and S90 samples, respectively. With these measurements, it was possible to observe that the 5A dye molecules are oriented in the same direction of the liquid crystal [21, 22]. An important fact is that S45 and S90 samples did not present the same line shape for the polarized absorption optical spectrum when the light interacted firstly in the different faces of samples. This clearly shows an asymmetry in the molecular ordering along the optical way, induced by liquid crystal molecular alignment, the exception in this case being the S00 sample. As a result, there are changes in the light polarization transmitted by this type of system [23].

**Table 1** shows $S_1$, $S_2$ and $S_3$ parameters normalized by $S_0$ obtained during EE measurements and its polarizations. In general, data of **Table 1** show that the change in the excitation wavelength does not change either the Stokes parameter and the degree polarization for each sample dramatically, mainly at 457 and 488 nm, due to absorption intensity (**Figure 4a**) is very close ($A_{457nm}/A_{514nm} = 0.91$). This means that the average orientation of molecules closer to the surface is what contributes the most to the emission. In the S00 sample, the emitted light is LPV due to the high value of $S_1$ with a negative signal. This is due to molecular alignment at 0° direction (or parallel to the y axis), and consequently, the LPV light is emitted. The high value of $S_2$ module shows that molecules are not fully oriented to the vertical direction. For this reason, a linear +45° polarized light is observed. This type of polarization does not change significantly in relation to the wavelength and polarization of excitation source. This result is corroborated by the different value of zero in the OA for perpendicular direction as shown in

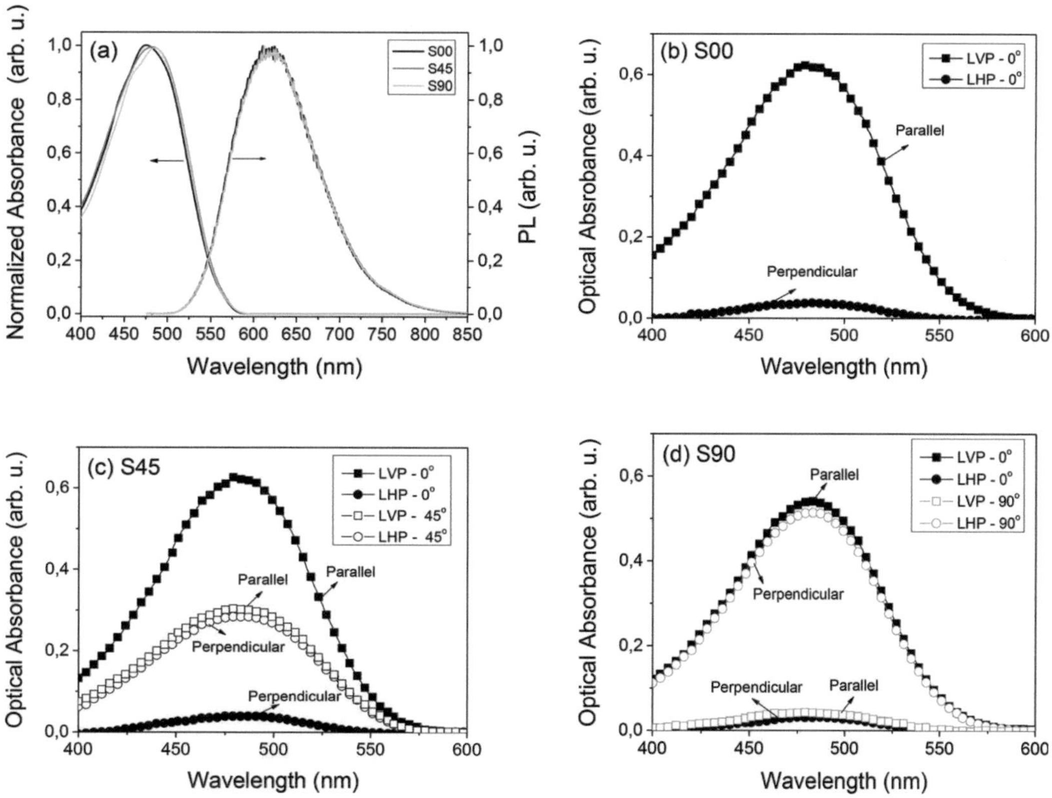


**Figure 4.** (a) OA and PL (excitation light at 457 nm) spectrums of S00, S45, and S90 samples. POA in parallel (vertical) and perpendicular (horizontal) direction of S00 (b), S45(c), and S90 samples (d). The direction at 0° of PVA films is considered parallel in relation to the vertical direction of laboratory.

**Figure 3b**. For the S45 sample, it can be seen by the $S_1$ value that on the face of 0° alignment, the light emitted is mostly linear vertically polarized as in the S00 sample. However, for the face with 45° alignment, this parameter decreases significantly. On the other hand, that face presents a significant increase in the $S_2$ parameter, which shows the emission of polarized light linear at 45° caused by the molecular ordering at 45°. Still for the S45 sample, the change in the wavelength and polarization of the excitation source does not cause a significant change in the Stokes parameter. Therefore, the emitted light remains with the same type of polarization. For the S90 sample, the light interacts first in the face with 0° alignment, independently of polarization and wavelength of excitation source, the emitted light is LVP as in the other samples. However, in the face with 90° alignment, the sample exhibits the emission of LHP light because the signal of $S_1$ is positive. This was expected due to molecules that, on this face, are horizontally oriented in relation to referential. An important point is that in the face with 0° alignment, this sample does not show linear polarized light at 45°, whereas on the face with 90° alignment, the emitted light presents a percentage of linear 45° polarized light. For the same face, when the wavelength of excitation source is at 514 nm, there is emission of circularly polarized light. This shows that molecules are rotating inside the liquid crystal cell and the emission of

| | | | $S_1/S_0$ | $S_2/S_0$ | $S_3/S_0$ | *P* | *r* |
|---|---|---|---|---|---|---|---|
| **Sample S00** | | | | | | | |
| **$\lambda_{exc}$ = 457 nm** | RCP | Face 00° | −0.67 | −0.33 | 0.09 | 0.75 | 0.58 |
| | LVP | Face 00° | −0.74 | −0.41 | 0.09 | 0.86 | 0.65 |
| **$\lambda_{exc}$ = 488 nm** | RCP | Face 00° | −0.55 | −0.33 | 0.07 | 0.68 | 0.45 |
| | LVP | Face 00° | −0.64 | −0.35 | 0.09 | 0.73 | 0.54 |
| **$\lambda_{exc}$ = 514 nm** | RCP | Face 00° | −0.47 | −0.47 | 0.05 | 0.67 | 0.37 |
| | LVP | Face 00° | −0.52 | −0.40 | 0.10 | 0.67 | 0.41 |
| **Sample S45** | | | | | | | |
| **$\lambda$exc = 457 nm** | RCP | Face 00° | −0.85 | −0.11 | 0.03 | 0.86 | 0.79 |
| | | Face 45° | −0.25 | 0.73 | 0.02 | 0.78 | 0.18 |
| | LVP | Face 00° | −0.84 | −0.10 | 0.05 | 0.85 | 0.77 |
| | | Face 45° | −0.25 | 0.76 | 0.04 | 0.81 | 0.18 |
| **$\lambda$exc = 488 nm** | RCP | Face 00° | −0.86 | −0.12 | 0.04 | 0.87 | 0.80 |
| | | Face 45° | −0.31 | 0.87 | 0.03 | 0.92 | 0.23 |
| | LVP | Face 00° | −0.87 | −0.12 | 0.03 | 0.88 | 0.81 |
| | | Face 45° | −0.32 | 0.86 | 0.03 | 0.92 | 0.23 |
| **$\lambda$exc = 514 nm** | RCP | Face 00° | −0.80 | −0.16 | 0.03 | 0.81 | 0.73 |
| | | Face 45° | −0.34 | 0.83 | 0.02 | 0.90 | 0.26 |
| | LVP | Face 00° | −0.85 | −0.18 | 0.13 | 0.88 | 0.79 |
| | | Face 45° | −0.32 | 0.82 | 0.02 | 0.89 | 0.23 |
| **Sample S90** | | | | | | | |
| **$\lambda$exc = 457 nm** | RCP | Face 00° | −0.79 | 0.03 | −0.06 | 0.79 | 0.71 |
| | | Face 90° | 0.80 | 0.20 | −0.05 | 0.82 | −0.42 |
| | LVP | Face 00° | −0.74 | 0.06 | 0.03 | 0.74 | 0.65 |
| | | Face 90° | 0.53 | 0.16 | −0.04 | 0.55 | −0.30 |
| **$\lambda$exc = 488 nm** | RCP | Face 00° | −0.80 | 0.03 | 0.02 | 0.80 | 0.73 |
| | | Face 90° | 0.74 | 0.16 | −0.07 | 0.76 | −0.40 |
| | LVP | Face 00° | −0.85 | 0.03 | 0.00 | 0.86 | 0.79 |
| | | Face 90° | 0.58 | 0.11 | −0.10 | 0.60 | −0.32 |
| **$\lambda$exc = 514 nm** | RCP | Face 00° | −0.85 | −0.04 | 0.12 | 0.86 | 0.79 |
| | | Face 90° | 0.78 | 0.20 | −0.12 | 0.81 | −0.41 |
| | LVP | Face 00° | −0.81 | −0.03 | 0.01 | 0.81 | 0.73 |
| | | Face 90° | 0.56 | 0.12 | −0.11 | 0.59 | −0.31 |

**Table 1.** Normalized Stokes parameter and polarization degree of S00, S45, and S90 samples.

LHP light occurs by molecules that are preferably oriented in the horizontal direction (**Figure 2d**), but have components in the vertical direction. For this reason, there is linear horizontally polarized emitted light even when the excitation source is LVP. The circularly polarized light can be easily observed when the excitation wavelength is at 514 nm. This happens because the length of penetration in the sample for this wavelength is higher ($A_{457nm}/A_{514nm}$ = 1.29) than the other wavelengths.

**Figure 5** shows two EE measurements performed in the S45 sample using the excitation light at 488 nm and LVP. By the fitting of EE data and using Eq. (18), Stokes parameters of samples' emitted light in function of the different excitation wavelengths and the type of the light polarization could be obtained. These measurements were performed on the two faces of the cell. Thereby, the molecular orientation within the liquid crystal cell by the emission of the dyes could be seen.

In order to verify the transmitted light properties, TE measurements were made at two wavelengths: 457 and 633 nm. Optical absorption of the 5A dye is present at 457 nm. For 633 nm changes in the polarization of transmitted light are caused only by the refractive index of the system, as there is not absorption. **Figure 6** shows two TSE measurements of the S45 and S90 samples.

By TE measurements, Stokes parameters were obtained and the polarization state of transmitted light was described. **Table 2** shows the Stokes parameters and the polarization degree of samples' transmitted light.

The first important point to discuss in these results is the polarization values higher than one. This occurs in three measures and it is due to the measurements' error that is around 5%. However, these values are not higher than recommended. From TE measurements, it could be seen that samples change the polarization of transmitted light and that this variation occurs even when the wavelength is not in the OA spectrum range of 5A dye. This is due to system birefringence [23]. Based on **Table 2**, it is clear that for the S00 sample, when the incident light

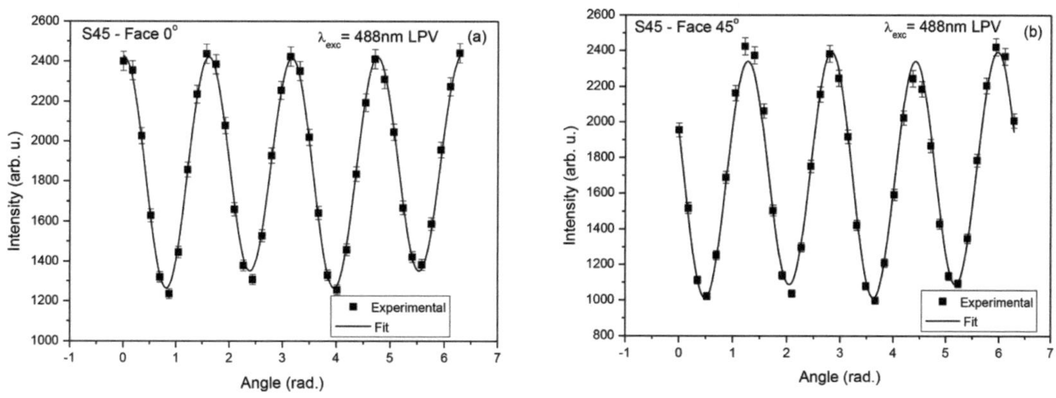


**Figure 5.** EE Measurement made in the S45 sample at face 0° (a) and 45° (b).

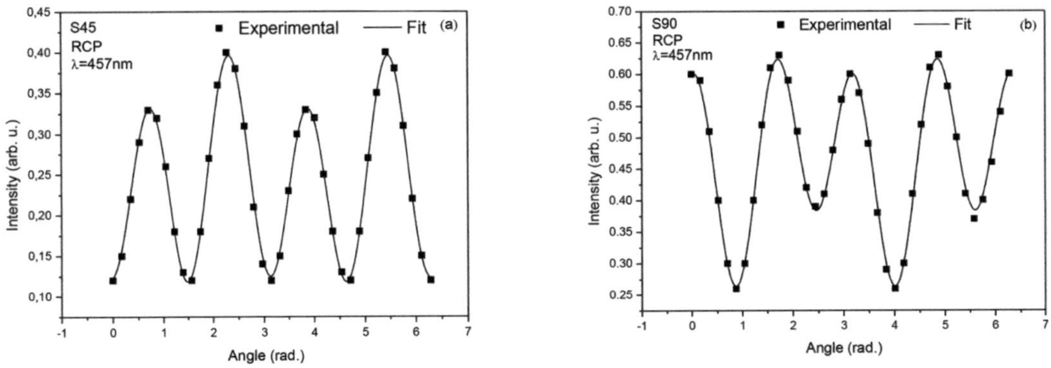


**Figure 6.** (a) Transmittance Ellipsometry measurements of the S45 sample, the light interacting firstly in the face with molecules oriented at 45° and the source at 457 nm is right circularly polarized. (b) Transmittance Ellipsometry measurements of the S90 sample, the light interacting firstly in the face with molecules oriented at 0° and the source at 457 nm is right circularly polarized.

is at 457 nm and right circularly polarized, the transmitted light is linear horizontally polarized due to the high value of $S_1$ with positive signal. There is also a significant amount of linear +45° polarization because of the value of $S_2$ and a small percentage of light is left circularly polarized. It is important to observe that the incident light is right circularly polarized, and the S00 sample changes the type of polarization of transmitted light for the small percentage of left circular polarization. In case of the incident light is left circularly polarized, the transmitted light through the sample is linear +45° polarized by the high value of $S_2$. There is also an amount of the linear horizontally polarized light due to the value of $S_1$, the transmitted light presents a small percentage of right circular polarization. When the incident light is linear vertically polarized, the transmitted light is linear vertically polarized by the $S_1$ value, and there is a small part of transmitted light that is right circularly polarized. Therefore, in this case, there are practically no changes in the polarization of incident light. Still for the S00 sample, when the incident light is right circularly polarized and the wavelength at 633 nm, the transmitted light changes its polarization for the linear −45°. There is a small percentage of light that is linear horizontally and right circularly polarized. For the measurement performed with the left circularly polarized incident light, a similar behavior of that when the incident light is right circularly polarized was shown. However, the transmitted light is linear +45° polarized, which is the majority of transmitted light polarization. When the incident light is linear vertically polarized, the sample does not change the polarization of transmitted light, which occurs in the same way when the wavelength is at 457nm and there is not optical absorption of dyes.

For the S45 sample, the same measurements as those of the S00 were obtained. However, in this sample, the incidence of light was carried out in the two faces of the sample, and it allows the verification of whether the polarization of transmitted light changes with molecular ordering. When the incident light has right circular polarization and wavelength at 457 nm, it can be observed that when the light interacts first in the face with 0° alignment, the transmitted light is linear +45° polarized, and a part of light is linear horizontally polarized, which should be

http://dx.doi.org/10.5772/intechopen.70089

| | | | $S_1/S_0$ | $S_2/S_0$ | $S_3/S_0$ | *P* |
|---|---|---|---|---|---|---|
| **Sample S00** | | | | | | |
| **$\lambda$ = 457 nm** | RCP | Face 00° | 0.62 | 0.36 | −0.16 | 0.74 |
| | LCP | Face 00° | 0.44 | −0.82 | 0.20 | 0.95 |
| | LVP | Face 00° | −1.00 | 0.03 | 0.13 | 1.01 |
| **$\lambda$ = 633 nm** | RCP | Face 00° | 0.11 | −0.79 | 0.09 | 0.80 |
| | LCP | Face 00° | −0.08 | 0.79 | 0.06 | 0.80 |
| | LVP | Face 00° | −0.95 | 0.06 | −0.03 | 0.95 |
| **Sample S45** | | | | | | |
| **$\lambda$ =457 nm** | RCP | Face 00° | 0.19 | 0.53 | 0.02 | 0.56 |
| | | Face 45° | 0.66 | −0.13 | 0.08 | 0.68 |
| | LCP | Face 00° | 0.05 | 0.63 | 0.03 | 0.63 |
| | | Face 45° | 0.69 | −0.02 | 0.08 | 0.69 |
| | LVP | Face 00° | −0.38 | −0.93 | 0.00 | 1.01 |
| | | Face 45° | 0.48 | −0.12 | −0.03 | 0.50 |
| **$\lambda$ = 633 nm** | RCP | Face 00° | 0.57 | 0.04 | 0.11 | 0.59 |
| | | Face 45° | 0.30 | −0.52 | 0.03 | 0.60 |
| | LCP | Face 00° | −0.22 | 0.03 | 0.45 | 0.50 |
| | | Face 45° | 0.22 | 0.11 | 0.55 | 0.61 |
| | LVP | Face 00° | −0.35 | −0.97 | 0.00 | 1.04 |
| | | Face 45° | −0.02 | −0.27 | −0.14 | 0.30 |
| **Sample S90** | | | | | | |
| **$\lambda$ = 457 nm** | RCP | Face 00° | −0.82 | −0.32 | 0.18 | 0.90 |
| | | Face 90° | 0.56 | 0.52 | −0.15 | 0.78 |
| | LCP | Face 00° | −0.53 | 0.37 | −0.52 | 0.84 |
| | | Face 90° | 0.63 | −0.39 | 0.42 | 0.85 |
| | LVP | Face 00° | 0.94 | −0.14 | 0.17 | 0.97 |
| | | Face 90° | 0.98 | −0.02 | −0.01 | 0.98 |
| **$\lambda$ = 633 nm** | RCP | Face 00° | −0.02 | 0.84 | 0.05 | 0.84 |
| | | Face 90° | 0.44 | −0.67 | 0.23 | 0.84 |
| | LCP | Face 00° | −0.37 | −0.68 | −0.42 | 0.88 |
| | | Face 90° | 0.24 | 0.80 | 0.00 | 0.84 |
| | LVP | Face 00° | 0.87 | −0.42 | −0.25 | 1.00 |
| | | Face 90° | 0.98 | −0.12 | −0.07 | 0.99 |

**Table 2.** Stokes parameters and the polarization degree of S00, S45, and S90 samples.

noticed for the $S_1$ value. This value increases significantly when the incidence occurs in the other face. On the other hand, the amount of linear at 45° polarized light decreases and changes the signal for negative. When the incident light is left circularly polarized, the transmitted light through the sample has the same state of that when the incidence occurs with the right circularly polarized light. However, the smallest part of polarized light ($S_1$ for the face at 0° and $S_2$ for the face at 45°) becomes practically zero. When the incident light is linear vertically polarized and the light interacts firstly with the face with 0° alignment, the polarization of transmitted light was changed to linear −45° polarization, there is still a small percentage of linear vertically polarized light. When the incidence occurs firstly in the face with 45° alignment, the transmitted light has linear horizontal polarization. However, the value of linear 45° polarized light that in the other face (0° alignment) is very high, decreases significantly. Measurements were made with the wavelength at 633 nm, and in this case, there is not OA of dye. The incident light is right circularly polarized and interacting first in the face with 0° alignment, the transmitted light has the linear horizontal polarization, and a small part of light is right circularly polarized. Nevertheless, when the light interacts first in the face with 45° alignment, the transmitted light is linear −45° polarized, there is still a part of the transmitted light that has the linear horizontal polarization. When the incident light is left circularly polarized and interacts first in the face with 0° alignment, the transmitted light has right circular polarization. There is still a part of the transmitted light that is linear vertically polarized. In the case of light interacting firstly with the other face (45° alignment), it can be verified that the majority of transmitted light is right circularly polarized, There is still a significant part of linear horizontally polarized light and a small part of linear +45° polarized light. When the incident light is linear vertically polarized, it can be observed that when the light interacts initially with the face with 0° alignment, the greater part of transmitted light has linear +45° polarization. There is also a percentage of linear vertically polarized light for the measurement performed when the light interacts first in the other face (45° alignment) of the sample. The transmitted light has very little polarization and the main polarization is linear −45°.

In the S90 sample, measurements were made in the same way as those of the S45 sample. For the incident light with wavelength at 457 nm and the right circular polarization, it can be seen that when the light interacts firstly with the face with 0° alignment, the transmitted light is linear vertically polarized. There is still a part of the light that is linear −45° polarized and a small part that is right circularly polarized. When the incident light interacts initially with the other face (90° alignment), the transmitted light has linear horizontal polarization, linear +45°, and a small part of light is left circularly polarized. When the incident light has left circular polarization, it could be verified that when the light interacts first in the face with 0° alignment, the transmitted light has a significant value of linear vertically and left circularly polarized light. There is also a small amount of linear +45° polarized light. When the light interacts initially with the other face (90° alignment), the transmitted light changes the signal of Stokes parameters in relation to the face with 0° alignment. Moreover, the transmitted light has linear horizontal polarization and it also has right circular polarization and a small percentage of linear −45° polarized light. When the incident light has linear vertical polarization, the transmitted light is practically linear horizontally polarized. There is only a small part of linear −45°

and right circularly polarized light when the incidence is performed firstly in the face with 0° alignment.

Like the other samples, measurements in the S90 were also made with the wavelength at 633 nm. When the incident light has right circular polarization, if the incident light interacts firstly in the face with 0° alignment, the transmitted light is linear +45° polarized. If the incidence occurs firstly in the other face (90° alignment), the transmitted light is mainly linear +45° polarized, in this case there is still a small amount of linear horizontal polarization and right circular polarization. When the incident light has left circular polarization, what happens is a similar process when the incident light is right circularly polarized? However, there is an inversion of the main polarization state of light between the face with 0° and 90° alignment. Measurement was also made with the incidence of linear vertically polarized light. For the two faces, the transmitted light is linear horizontally polarized. Therefore, the transmitted light changes completely its type of polarization. It is possible to observe that when the incident light interacts initially in the face with 0° alignment, the transmitted light has a percentage of linear −45° and left circularly polarized light. For the incidence performed in the other face (90° alignment) a percentage of this polarization state for the transmitted light is much smaller in relation to the incidence achieved firstly in the face with 0° alignment.

With EE and TE measurements, a complete molecular ordering inside the doped liquid crystal cell could be verified. It is important to observe that the EE measurements, in general, allow characterization of the molecular ordering onto the substrate surface. It can be changed by the excitation wavelength and molecules along the cell. With TE measurements, results allow the presence of an effect that occurs in the molecular ordering along of the doped liquid crystal cell. This can be related to the birefringence of system only when the light is transmitted by the sample and there is no optical absorption of 5A dye, or it can be related to the optical absorption and the birefringence simultaneously.

## 4. Conclusion

In this work, we illustrated a new methodology for measuring Raman Optical Activity [14] and the application of achromatic Ellipsometry in the study of benzotiadiazol derivative dye doped in the E7 nematic liquid crystal. Emission and transmittance Ellipsometry experiments were used to verify its molecular ordering within a liquid crystal cell. The obtained results showed that chromophores were ordered in the same direction of the E7 molecules. In addition, due to the change in the excitation wavelength, penetration depth in the sample could be verified, enabling the analysis of molecules ordering within a liquid crystal cell. Ellipsometry experiments also indicate that the molecular orientation inside the cells somehow changes the polarization of transmitted light. This effect can be due to system birefringence (both liquid crystal and dye) or due to combined process between dyes optical absorption and system birefringence. It is important to emphasize that changes in light polarization occur differently for one process (birefringence) or two processes simultaneously (birefringence and optical absorption). Finally, achromatic Ellipsometry was demonstrated to be a complete spectroscopic technique that enables the analysis of bulk and interfacial molecular ordering.

## Acknowledgements

The authors are grateful to the following Brazilian Agencies: FAPEMAT, FAPEMIG, CNPQ, INCT/INEO, and INCT/INFo. The authors are grateful to Heidi C. Piva for helping with grammar corrections.

## Author details

Eralci Moreira Therézio[1]*, Gustavo G. Dalkiranis[2], André A. Vieira[3], Hugo Gallardo[4], Ivan H. Bechtold[5], Patricia Targon Campana[6] and Alexandre Marletta[7]

*Address all correspondence to: therezio@ufmt.br

1 Institute of Exact and Natural Sciences, Federal University of Mato Grosso, Rondonópolis, Brazil

2 Physics Department, Universitat Autònoma de Barcelona, Bellaterra, Spain

3 Chemistry Institute, Federal University of Bahia, Salvador, Brazil

4 Chemistry Department, Federal University of Santa Catarina, Florianópolis, Brazil

5 Physics Department, Federal University of Santa Catarina, Florianópolis, Brazil

6 School of Arts, Sciences and Humanities, University of São Paulo, São Paulo, Brazil

7 Physics Institute, Federal University of Uberlândia, Uberlândia, Brazil

## References

[1] Marletta A. Optical Properties of Organic Semiconductors Based on Light Emitting Polymers. São Carlos: Universidade de São Paulo; 2001

[2] Alliprandini-Filho P. Emission Ellipsometry Technique Application for Luminescent Materials Characterization. Uberlândia: Universidade Federal de Uberlândia; 2012

[3] Valeur B. Molecular Fluorescence: Principles and Applications. Weinheim: Wiley-VCH; 2001

[4] Kaito A, Nakayama K, Kanetsuna H. Infrared dichroism and visible-ultraviolet dichroism studies on roller-drawn polypropylene and polyethylene sheets. Journal of Macromolecular Science, Part B. 1987;**26**(3):281–306

[5] Alliprandini-Filho P, da Silva GB, Barbosa Neto NM, Silva RA, Marletta A. Induced secondary structure in nanostructured films of poly(p-phenylene vinylene). Journal of Nanoscience and Nanotechnology. 2009;**9**:5981–5989

[6] Berova N, Nakanishi K, Woody RW. Circular Dichroism: Principles and Applications. New York: Wiley-VCH; 1994

[7] Lin SH. Band shape of the circular dichroism. The Journal of Chemical Physics. 1971;**55** (7):3546–3554

[8] Shirvani-Mahdavi H, Mohajerani E, Wu S-T. Circularly polarized high-efficiency cholesteric liquid crystal lasers with a tunable nematic phase retarder. Optics Express. 2009;**18** (5):5021–5027

[9] Oda M, Nothofer HG, Lieser G, Scherf U, Meskers SCJ, Neher D. Circularly polarized electroluminescence from liquid-crystalline chiral polyfluorenes. Advanced Materials. 2000;**12**(5):362–365

[10] Therézio EM, Piovesan E, Vega ML, Silva RA, Oliveira ON, Marletta A. Thickness and annealing temperature effects on the optical properties and surface morphology of layer-by-layer poly(p-phenyline vinylene)+dodecylbenzenesulfonate films. Journal of Polymer Science Part B: Polymer Physics. 2011;**49**(3):206–213

[11] Alliprandini-Filho P, da Silva RA, Barbosa Neto NM, Marletta A. Partially polarized fluorescence emitted by MEHPPV in solution. Chemical Physics Letters. 2009;**469**(1-3):94–98

[12] Collet E. Polarized Light: Fundamentals and Applications. New York: Marcel Dekker; 1993

[13] Therézio EM, da Silva SFC, Dalkiranis GG, Alliprandini-Filho P, Santos GC, Ely F, et al. Light polarization states of a cholesteric liquid crystal probed with optical ellipsometry. Optical Materials. 2015;**48**:7–11

[14] Basílio FC, Campana PT, Therézio EM, Barbosa Neto NM, Serein-Spirau F, Silva RA, et al. Ellipsometric Raman spectroscopy. The Journal of Physical Chemistry C. 2016;**120** (43):25101–25109

[15] Basílio FC. Implementation of the New Technique Raman Spectroscopy by Elipsometry no Chiral Molecules Study. Uberlândia: Universidade Federal de Uberlândia; 2014

[16] Stokes GG. On the composition and resolution of streams of polarized light from different sources. Transactions of the Cambridge Philosophical Society. 1852;**9**:399–416

[17] Therézio EM, Franchello F, Dias IFL, Laureto E, Foschini M, Bottecchia OL, et al. Emission ellipsometry as a tool for optimizing the electrosynthesis of conjugated polymers thin films. Thin Solid Films. 2013;**527**:255–260

[18] Therézio EM, Rodrigues PC, Tozoni JR, Marletta A, Akcelrud L. Energy-transfer processes in donor-acceptor poly(fluorenevinylene-alt-4,7-dithienyl-2,1,3-benzothiadiazole). The Journal of Physical Chemistry C. 2013;**117**(25):13173–13180

[19] Pereira GGD. Study of liquid crystal mesophases by emission ellipsometry technique [dissertation]. Uberlândia: Universidade Federal de Uberlândia; 2013

[20] Alliprandini-Filho P, Dalkiranis GG, Armond RASZ, Therezio EM, Bechtold IH, Vieira AA, et al. Emission ellipsometry used to probe aggregation of the luminescent 2,1,3-benzothiadiazole dyes and ordering in an E7 liquid crystal matrix. Physical Chemistry Chemical Physics. 2014;**16**(7):2892–2896

[21] Galyametdinov YG, Knyazev AA, Dzhabarov VI, Cardinaels T, Driesen K, Görller-Walrand C, et al. Polarized luminescence from aligned samples of nematogenic lanthanide complexes. Advanced Materials. 2008;**20**(2):252–257

[22] Alliprandini-Filho P, Borges GF, Calixto WB, Bechtold IH, Vieira AA, Cristiano R, et al. Molecular alignment effects on spectroscopic properties 2,1,3-benzothiadiazole guested in liquid–crystalline compounds. Chemical Physics Letters. 2010;**487**(4-6):263–267

[23] Cristaldi DJR, Pennisi S, Pulvirenti F. Liquid Crystal Display Drivers: Techniques and Circuits. Catania: Springer; 2009

Chapter 3

# The Development of Three-Intensity Measurement in PSA Ellipsometry and Photoelastic Modulation Ellipsometry

Yu-Faye Chao

Additional information is available at the end of the chapter

http://dx.doi.org/10.5772/intechopen.70086

**Abstract**

A three-intensity measurement technique in a polarizer-sample-analyzer (PSA) ellipsometry will be introduced. The alignment of the azimuth angle of polarizer and analyzer to the incident plane will be discussed. Its applications to measure the ellipsometric parameters for deducing the optical parameters will be stated. In addition to the PSA ellipsometry, one can insert a photoelastic modulator (PEM) in the PSA ellipsometry for developing a PEM ellipsometry. There is no moving part in the system and its measuring speed is only limited by a modulator. An in situ/real-time and post-flight ellipsometry can be established for monitoring the dynamically varying process.

**Keywords:** alignment, calibrations, curved surface, real time

## 1. Introduction

The ellipsometry measures the change in polarization upon reflection or transmission [1] and then compares it with a model to deduce the optical parameters, such as the complex refractive index, film thickness, crystalline nature, and roughness of a sample. The ratio of the reflection coefficient in p-polarized light ($r_p$) to s-polarized light ($r_s$) has been noted as $\tan\Psi\, e^{i\Delta}$, where $\Psi$ and $\Delta$ are the two ellipsometric parameters measured in ellipsometer. These parameters are very sensitive to the change in the optical response of incident radiation that interacts with the material. A commercial ellipsometer for measuring the thickness of a thin film has to achieve the precision within 2 Å in order to attract the IC industry for thin film measurement. This means the precision of $\Psi$ and $\Delta$ have to be less than 0.02° and 0.04°, respectively. In ellipsometry, the alignment and calibration are the crucial techniques for achieving these precisions. These techniques fully depend on the configuration of the system. First, one has to establish the plane of incident by aligning the azimuth angle of polarizer and analyzer to the plane of incident, both of which are the fundamental components of ellipsometer. Collins [2] and Chao [3] applied the null technique and the intensity ratio technique to locate the azimuth zeros of polarizer (P) and analyzer (A) with respect to the plane of incident, respectively. Both methods are complicated and time consuming. Due to the development of symbolic computer program Mathematica, one can analytically obtain the zeros of polarizer and analyzer with respect to the incident plane. In this chapter, the azimuth zeros of P and A to the incident plane will be located by an intensity ratio method. By rotating the analyzer, a three-intensity measurement technique will be applied to measure the ellipsometric parameters. According to the theoretical models, one can deduce the optical parameters from the measured ellipsometric parameters. Since only three intensities are needed to be measured, the imaging ellipsometry can be established. By using the following models: (a) the film thickness of flat and curved surface; (b) the refractive indices of homogeneous and nonhomogeneous bulk medium; (c) focused beam for surface plasmon resonance (SPR) phenomena, one can prove that this simple imaging ellipsometry has many different applications.

The photoelastic modulator (PEM) will be introduced in the polarizer-sample-analyzer (PSA) ellipsometric system for an in situ/real-time ellipsometry. The in situ alignment technique of the optical axis of PEM is essential for establishing a real-time ellipsometry. Here, the real-time measurements of the ellipsometric parameters under etching and thermal heating processes will be presented. In addition to real time, one can also use the post-flight technique to analyze the digital waveform of an anisotropic medium.

## 2. The alignment in PSA ellipsometry

The PSA ellipsometry setup is shown in **Figure 1**. Before inserting the PEM, one can setup the PSA ellipsometry for three-intensity measurement. For aligning the azimuth angles of P and A with respect to the incident plane, one can setup the PSA configuration in the interested incident angle (typical 70° in IC industry). The measured reflected intensity is expressed in the ellipsometric parameters, Ψ, Δ, and azimuth angles of P and A:

$$I(P,A) = I_0\left[\sin^2 P \sin^2 A + \tan^2\Psi \cos^2 P \cos^2 A + \frac{1}{2}\tan\Psi \cos\Delta \sin 2P \sin 2A\right] \tag{1}$$

By considering the polarizer and analyzer separately, Chao [4] improved Steel's [5] intensity ratio technique to align the polarizer and analyzer to the incident plane. For alignment, the following intensity ratios A and B are defined:

$$A(\alpha,\beta) = \frac{I(45^\circ+\alpha,\beta)}{I(45^\circ+\alpha,90^\circ+\beta)} \tag{2}$$

$$B(\alpha,\beta) = \frac{I(-45^\circ+\alpha,\beta)}{I(-45^\circ+\alpha,90^\circ+\beta)} \tag{3}$$

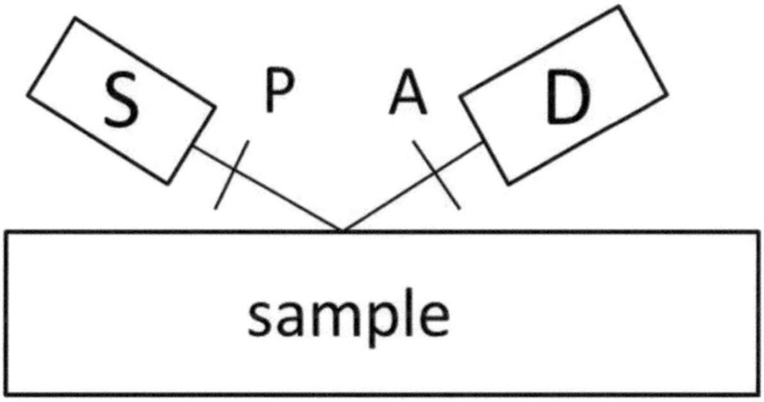


**Figure 1.** Schematic setup of PSA ellipsometry: A: analyzer, S, light source; P, polarizer; D, detector.

In addition to the azimuth angle of P and A, the small angle deviation $\alpha$ and $\beta$ from the incident plane, respectively, are considered. These intensity ratios are expanded to the first-order approximation around their respective scales, and expressed as follows:

$$A(\alpha,\beta) \cong \tan^2\psi - 2\left[2\tan^2\Psi\ \alpha - \sec^2\Psi\ \tan\Psi\ \cos\Delta\ \beta\right] \tag{4}$$

$$B(\alpha,\beta) \cong \tan^2\Psi + 2\left[2\tan^2\Psi\ \alpha - \sec^2\Psi\ \tan\Psi\ \cos\Delta\ \beta\right] \tag{5}$$

These two equations intersect with each other when

$$2\tan^2\Psi\ \alpha - \sec^2\Psi\ \tan\Psi\ \cos\Delta\ \beta = 0 \tag{6}$$

so the deviation of $\alpha$ and $\beta$ can be easily obtained for the opposite of $\cos\Delta$. Two incident angles [3], and fixed incident angle by two samples and two wavelengths [6] are proposed to align the azimuth angles of polarizer and analyzer. These complicated processes can be substituted by an analytical measurement technique [7]. In polar coordinates, if the incident light is $E_p$=$E_s$, the reflected intensity is distributed in an elliptical form [5] as shown in **Figure 2**, which can be formulated as

$$I(A) = L\cos^2(A-\theta) + T\sin^2(A-\theta) \tag{7}$$

where L and T are the magnitudes of the maximum and minimum intensities, respectively, and $\theta$ is the azimuth of maximum intensity. Since there are only three unknowns, one can measure three intensities to deduce the three parameters. Eq. (7) can also be written as follows:

$$I(A) = B(1 + C\cos 2A + D\sin 2A) \tag{8}$$

where the parameters B, C, and D can be written in terms of L, T, and $\theta$, that is, B = 0.5(L+T), C = (L−T)cos2$\theta$/(L+T), and D = (L−T)sin2$\theta$/(L+T). It is easy to prove that parameters B, C, and D can be measured by three intensities measured at A = 0°, 60°, and 120°;

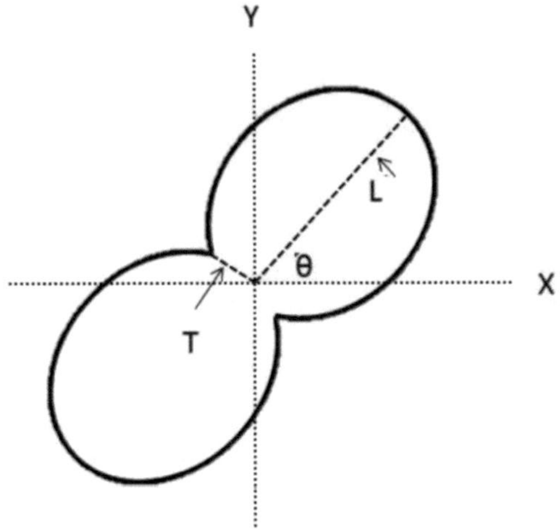


**Figure 2.** Intensity distribution in polar coordinates at various azimuth angles of analyzer.

$$B = \frac{1}{3}[I(0^\circ) + I(60^\circ) + I(120^\circ)] \tag{9}$$

$$C = 2 - [I(60^\circ) + I(120^\circ)]/B \tag{10}$$

$$D = \frac{1}{B\sqrt{3}}[I(60^\circ) - I(120^\circ)] \tag{11}$$

So, tan2θ = D/C; L = B+DB/sin2θ and T = 2B−L. By comparing the expansion of Eq. (1), one can obtain the following relationships:

$$\tan 2\theta = \frac{\cos\Delta \ \sin 2P \ \sin 2\Psi}{\cos 2P - \cos 2\Psi} \tag{12}$$

$$\tan^2\Psi = \frac{1+C}{1-C}\tan^2 P \tag{13}$$

$$\frac{(L-T)^2}{4LT}\sin^2 2\theta = \cot^2\Delta \tag{14}$$

when there is no deviation in the azimuth angles of polarizer and analyzer. Even there are deviations $\alpha$ and $\beta$ for P and A, respectively, one can substitute those errors in the previous relations and obtain the analytical results, that is

$$\sin 2\alpha = \frac{1 - \sqrt{\frac{(1+C_1)(1-C_2)}{(1-C_1)(1+C_2)}}}{1 + \sqrt{\frac{(1+C_1)(1-C_2)}{(1-C_1)(1+C_2)}}} \tag{15}$$

$$\tan\Psi = 4\sqrt{\frac{1 - \sqrt{\frac{(1+C_1)(1-C_2)}{(1-C_1)(1+C_2)}}}{1 + \sqrt{\frac{(1+C_1)(1-C_2)}{(1-C_1)(1+C_2)}}}} \tag{16}$$

$$\beta = \frac{1}{4}\left(\tan^{-1}\frac{D_1}{C_1} - \tan^{-1}\frac{D_2}{C_2}\right) \tag{17}$$

where subscripts 1 and 2 indicate the three-intensity measurement obtained at P = 45° and −45°, respectively. After correction, the value Δ can be obtained by substituting the corrected value into the appropriate relation. The most important aspect of this technique is the concept of direct determination, in contrast to locating the minimum intensity in the null ellipsometry, which not only requires a sensitivity detector but also requires a polarizer with a high extinction ratio. This intensity ratio technique only needs a minimum amount of data to deduce the deviations and ellipsometric parameters simultaneously. If the detector is a high-dynamic range detector, the precision of this three-intensity measurement technique will be compatible to the commercial ellipsometer. Moreover, a complicated alignment process is not required in the simple PSA ellipsometry.

## 3. Three-intensity measurements technique in PSA ellipsometry

The analytical alignment technique provides a simple and accurate alignment technique in ellipsometry. Since only three-intensity measurements are required, one can use charge-coupled device (CCD) as a detector for imaging ellipsometry.

### 3.1. Film thickness of $SiO_2$/Si thin film

In the beginning, the azimuth angles of polarizer and analyzer are aligned to the incident plane by the Brewster angle of a nonabsorbent as the reference zero. After the rough alignment, the azimuth angle of polarizer is set at 45° and −45°, respectively, for analyzer to be at three angles, that is, 0°, 60°, and 120°, and the corresponding three-intensity measurements. One can choose an appropriate incident angle (typical 70°) for ellipsometric measurements. Using the relations in the last section, one can calculate Ψ, Δ, α, and β from three-intensity measurements. According to the definition of ellipsometric parameters to Fresnel reflection coefficient [1], one can deduce the film thickness of thin film ($SiO_2$/Si thin film) through computer program. The primary objective is to compare the ellipsometric parameters measured with rough alignment to the analytically corrected values. Just as expected, the system errors are clearly observable in the deduced values of Ψ and Δ, measured at P = 45° and −45° separately, as shown in **Figure 3**. After readjusting the azimuth angles of polarizer and analyzer, one can observe the flat distribution of Ψ and Δ measured at P = 45° and −45°. The corrected ellipsometric parameters of the results deviated from each other by less than 0.01°, as shown in **Tables 1** and **2**. The corrected values are used for film thickness deduction. It is known that a rotating polarizer ellipsometry can deviate the incident angle [8]; the incident angle for thickness deduction is 69.94° instead of 70°.

### 3.2. Imaging ellipsometry

In the PSA ellipsometry, one can add a beam expander and use CCD as a detector, such as shown in **Figure 4**. This imaging ellipsometry can measure the refractive index of a flat bulk

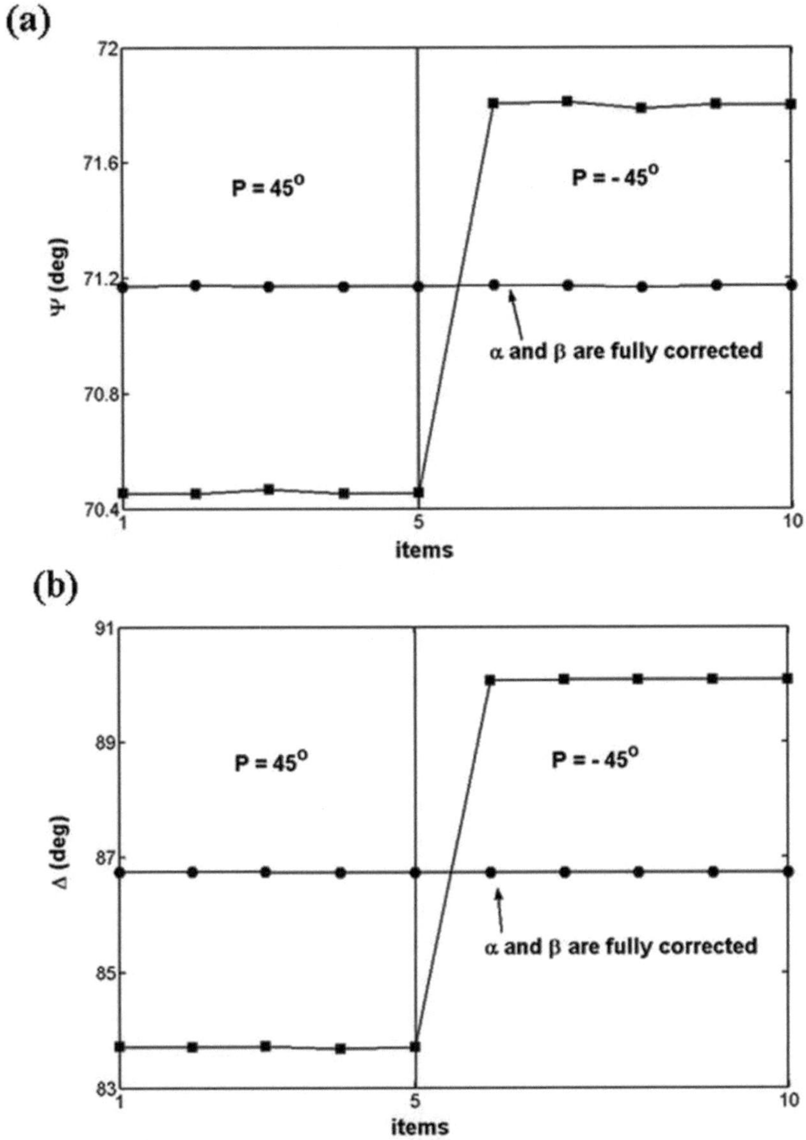


**Figure 3.** Ellipsometric measurements of PSA ellipsometry for a $SiO_2$ film on a Si substrate: (a) ψ and (b) Δ measured after a rough alignment (filled squares) at P = 45° and −45°, where $\alpha = 0.998\pm0.011°$ and $\beta = 1.233\pm0.005°$ and after readjusted (filled circles) at P = 45° and −45°, where $\alpha = 0.015\pm0.009°$ and $\beta = 0.006\pm0.003°$.

medium, curved surface, and film thickness on a curved surface. Moreover, the SPR phenomena will be presented by a focused light beam.

### *3.2.1. Refractive index of flat surface of a bulk medium*

Since there is no multiple reflection on a bulk medium, the refractive index of a bulk medium can be easily obtained by the ellipsometric parameter from the following relationship, let $\rho = \tan\Psi\, e^{i\Delta}$, then

http://dx.doi.org/10.5772/intechopen.70086

| Experiment | ψ (deg) | Δ (deg) | Thickness (nm) | α (deg) | β (deg) |
|---|---|---|---|---|---|
| 1 | 71.172 | 86.742 | 129.89 | 0.993 | −1.229 |
| 2 | 71.175 | 86.745 | 129.93 | 1.000 | −1.234 |
| 3 | 71.169 | 86.754 | 129.85 | 0.968 | −1.231 |
| 4 | 71.171 | 86.727 | 129.88 | 0.991 | −1.241 |
| Mean | 71.171 | 86.741 | 129.88 | 0.988 | −1.233 |
| Standard deviation | 0.003 | 0.011 | 0.03 | 0.011 | 0.005 |
| Jobin-Yvon ellipsometer | | | 129.3 | | |

**Table 1.** Optimized ellipsometric measurements of PSA ellipsometry for a $SiO_2$ film on a Si substrate after rough alignment.

| Experiment | ψ (deg) | Δ (deg) | Thickness (nm) | α (deg) | β (deg) |
|---|---|---|---|---|---|
| 1 | 71.176 | 86.739 | 129.94 | 0.017 | −0.011 |
| 2 | 71.174 | 86.731 | 129.93 | 0.013 | −0.005 |
| 3 | 71.166 | 86.728 | 129.87 | 0.003 | −0.005 |
| 4 | 71.174 | 86.726 | 129.91 | 0.026 | −0.004 |
| Mean | 71.173 | 86.731 | 129.91 | 0.015 | −0.006 |
| Standard deviation | 0.004 | 0.006 | 0.03 | 0.009 | 0.003 |

**Table 2.** Optimized ellipsometric measurements of PSA ellipsometry for a $SiO_2$ film on a Si substrate after readjusting azimuth angles of polarizer and analyzer.

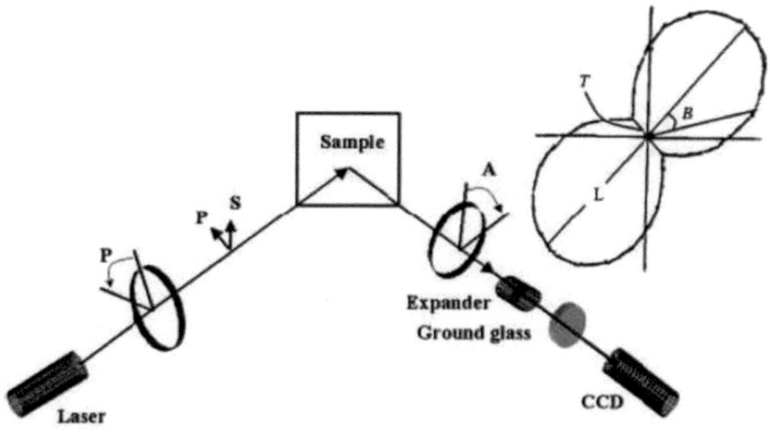


**Figure 4.** Schematic setup of the PSA imaging ellipsometer: Laser; P, polarizer; A, analyzer; S, sample; Expander, beam expander, and CCD camera. The elliptically distributed intensity on the right is under various azimuth angles of A in the polar coordinates.

$$n_1 = n_0 \tan \theta_i \left[ 1 - \frac{4\rho}{(1+\rho)^2} \sin^2 \theta_i \right]^{\frac{1}{2}} \tag{18}$$

where $n_1$, $n_0$ are the refractive indices of bulk medium and incident medium, respectively. θι is the incident angle.

The graded index structure has been widely used in fiber communication, in technologies such as the graded index fiber, and the gradient-index (GRIN) lens. The knowledge of the refractive index profile (RIP) of this graded index structure is important not only to assess its performance in optical devices but also to control the quality of products [9]. Since it is very important to measure the RIP of a gradient index material, the nondestructive measurement using imaging ellipsometry is suggested. The well-known material BK7 glass can be used to evaluate this imaging ellipsometry, such as shown in **Figure 5**. Since glass is nonabsorbent, the refractive index can be obtained by Ψ. In addition, the RIP of a gradient index fiber can be measured; its surface property can also be examined by α, such as shown in **Figure 6**, which will be proved that it is the direction of surface's normal.

*3.2.2. The curved surface*

Surface topography is a topic of great interest to science, technology, and industry. Both the contact stylus-based instruments [10] and noncontact optical instruments [11] can be used to measure and characterize the topography of a surface. By performing the ellipsometric measurements at two azimuth angles of the polarizer that differ by 90°, one can determine the azimuth deviation (α) of polarizer with respect to the incident plan by the three-intensity technique of PSA ellipsometry. After adjusting the deviation, one can prove that the azimuth

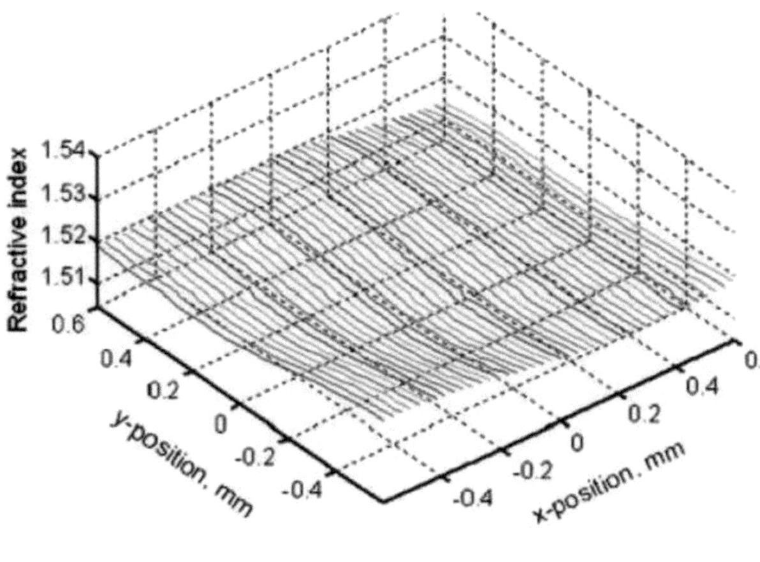


**Figure 5.** Imaging ellipsometric study of a BK7 glass: n = 1.520±0.003.

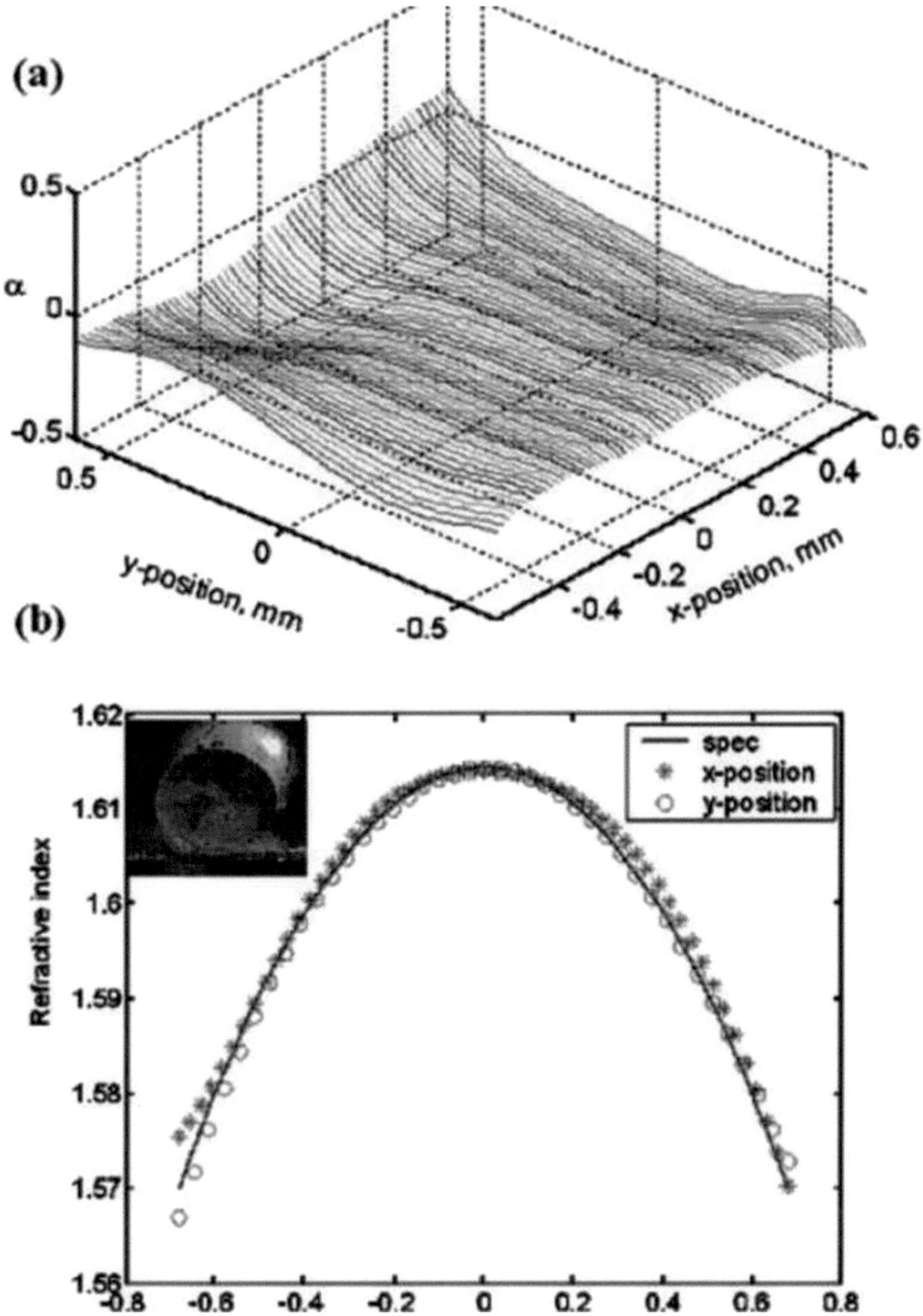


**Figure 6.** Imaging ellipsometric study of a flat GRIN lens: (a) azimuth deviation ($\alpha$) of the surface. (b) Measured refractive index profile (*: x-axis, O: y-axis, solid line: RIP provided by NSG).

deviation of polarizer measured by this imaging ellipsometry can be used to measure the surface's normal; thus, it can be used to measure the surface topography of a lens. In the beginning, the alcohol surface is regarded as a flat surface for calibration; then one can tilt an $SiO_2$/Si sample for investigating its effect to the azimuth deviation of polarizer. In the previous section, the refractive index profile of a GRIN lens has been obtained. Here, one can use the same technique to measure the RIP of the diluted alcohol, such as shown in **Figure 7**. Since both the GRIN lens and diluted alcohol have flat surfaces, their azimuth deviations are near to zero. Tilting a plate confirms that $\alpha$ can be used to measure the normal line of the surface, such as shown in **Figure 8**. After understanding the meaning of $\alpha$, one can measure the refractive index of a BK7 plano-convex lens and its surface's normal, such as shown in **Figure 9**. This can be flattened by correcting the $\alpha$, as shown in **Figure 10**. Furthermore, one can rotate the tilted $SiO_2$/Si to investigate the three-direction cosines of the surface's normal. From the top view of a tilted plane, such as shown in **Figure 11b**, one can identify the surface's normal deviated by the tilted angle $\Theta$ with components $\alpha$, and $\Delta\theta$ along x- and y-axes, respectively. The measured deviation $\alpha$ is the surface's normal projected on x-axis, which is perpendicular to the intersection of incident plane and sample surface. $\Delta\theta$ is the deviation of the incident angle caused by the tilting and can be deduced from the three-direction cosines. Three positions are examined, **Table 3**; all these results can be used to deduce the thickness of the thin film $SiO_2$ on silicon substrate [12]. For simplicity, one shall use a cylindrical lens and set it up as in **Figure 12**, then examine the thickness of its coating by the three-intensity measurements technique, the

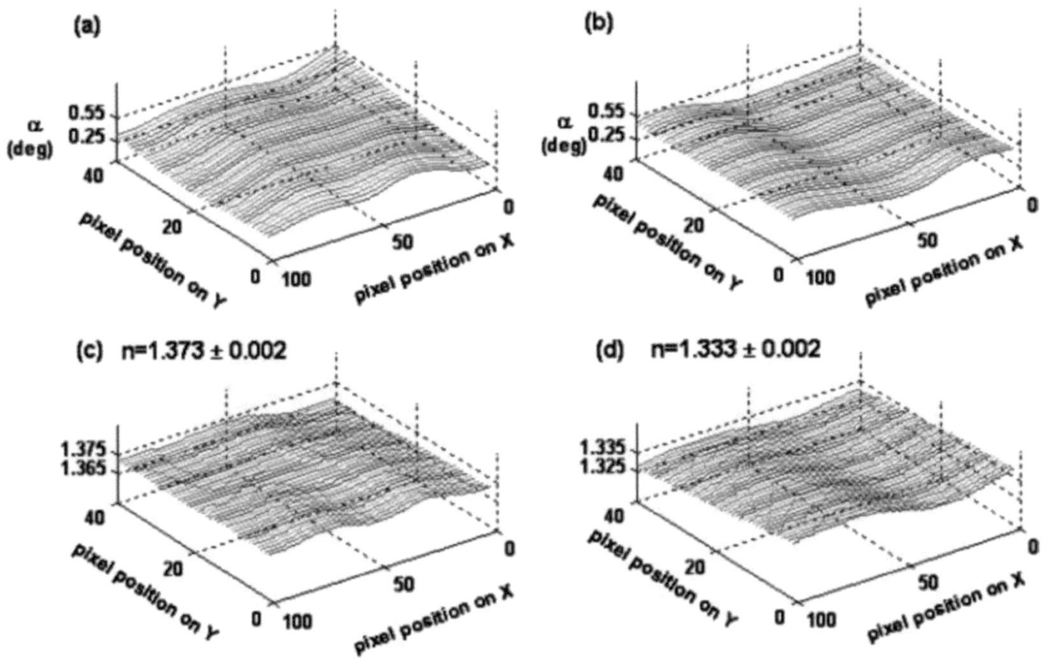


**Figure 7.** Imaging ellipsometric study of water and diluted alcohol: (a) azimuth deviation ($\alpha$) of the diluted alcohol surface, (b) azimuth deviation ($\alpha$) of the water surface, (c) refractive index profile of diluted alcohol n = 1.373 ± 0.002, and (d) refractive index profile of water n = 1.333 ± 0.002.

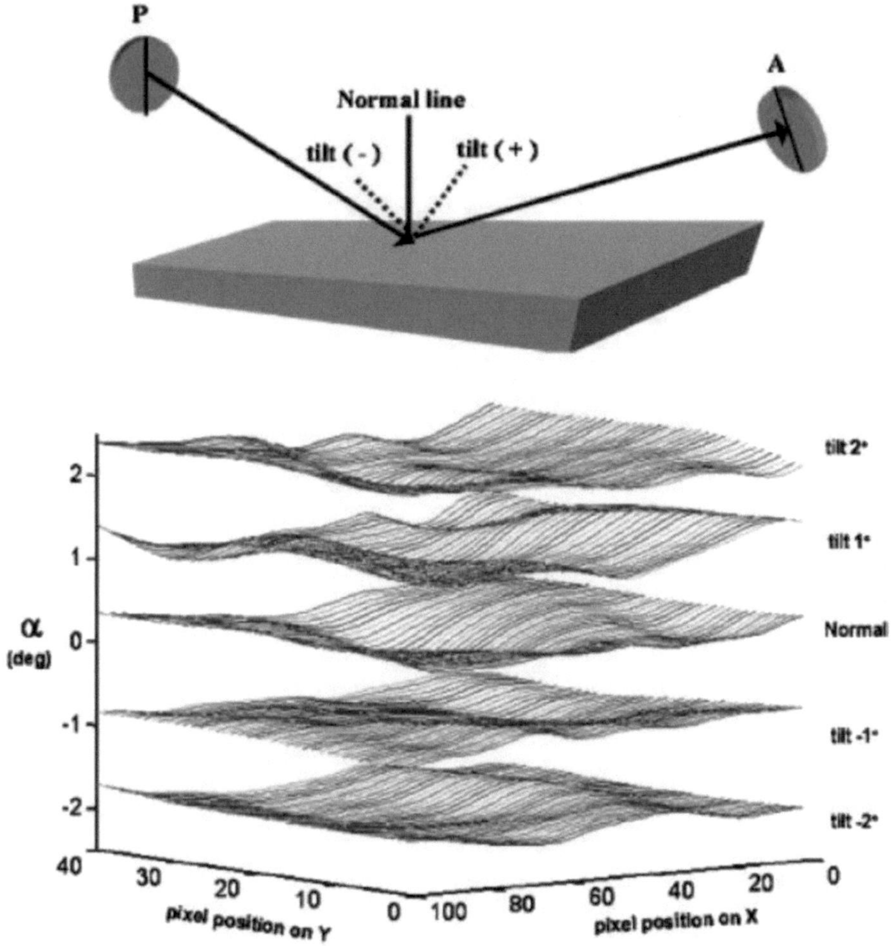


**Figure 8.** Imaging ellipsometric study of azimuth deviation of polarizer (a) caused by the tilting of a $SiO_2$ thin film on Si substrate. The thin film is tilted from −2 to 2° with step 1°.

measured Ψ and Δ are shown in **Figure 13**. In this configuration, the measured $\alpha$ is zero at the vertex of the cylinder. The corrected Ψ and Δ are flat which can be used to deduct the thickness of the coating. In conclusion, this three-intensity measurement technique can perform a measurement on thin film coating of a curved and titled surface with high precision.

### *3.2.3. Focused beam and surface plasmon resonance*

In conventional ellipsometry, the light beam has to be collimated carefully to avoid the polarization errors. In the last section, we show that one can perform the ellipsometric measurement of a curved surface. Since the three-direction cosines have been well understood, one can measure a convergent/ divergent beam without any restriction [13]. Using a focused beam in the PSA imaging ellipsometry, one can perform the ellipsometric measurement under the surface plasmon resonance (SPR) condition and observe the resonance phenomena.

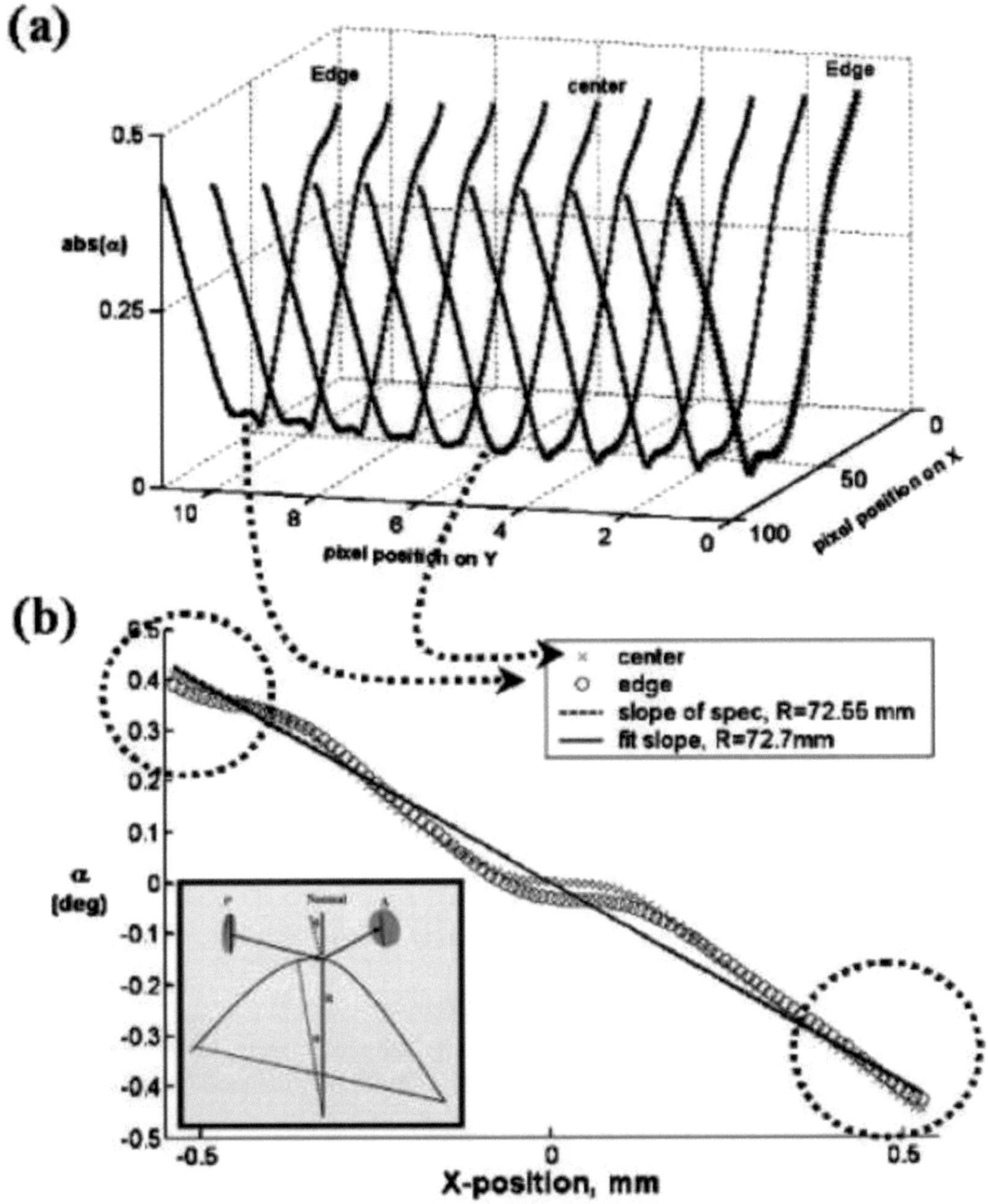


**Figure 9.** Imaging ellipsometric study of the radius of curvature of the BK7 plano-convex lens: (a) the magnitude of the azimuth deviation ($\alpha$) of the surface and (b) the derived and measured $\alpha$ of the lens at the center and the edge (dashed line: derived from the radius measured by the Zygo phase-shift interferometer, solid line: measured by the imaging ellipsometric technique).

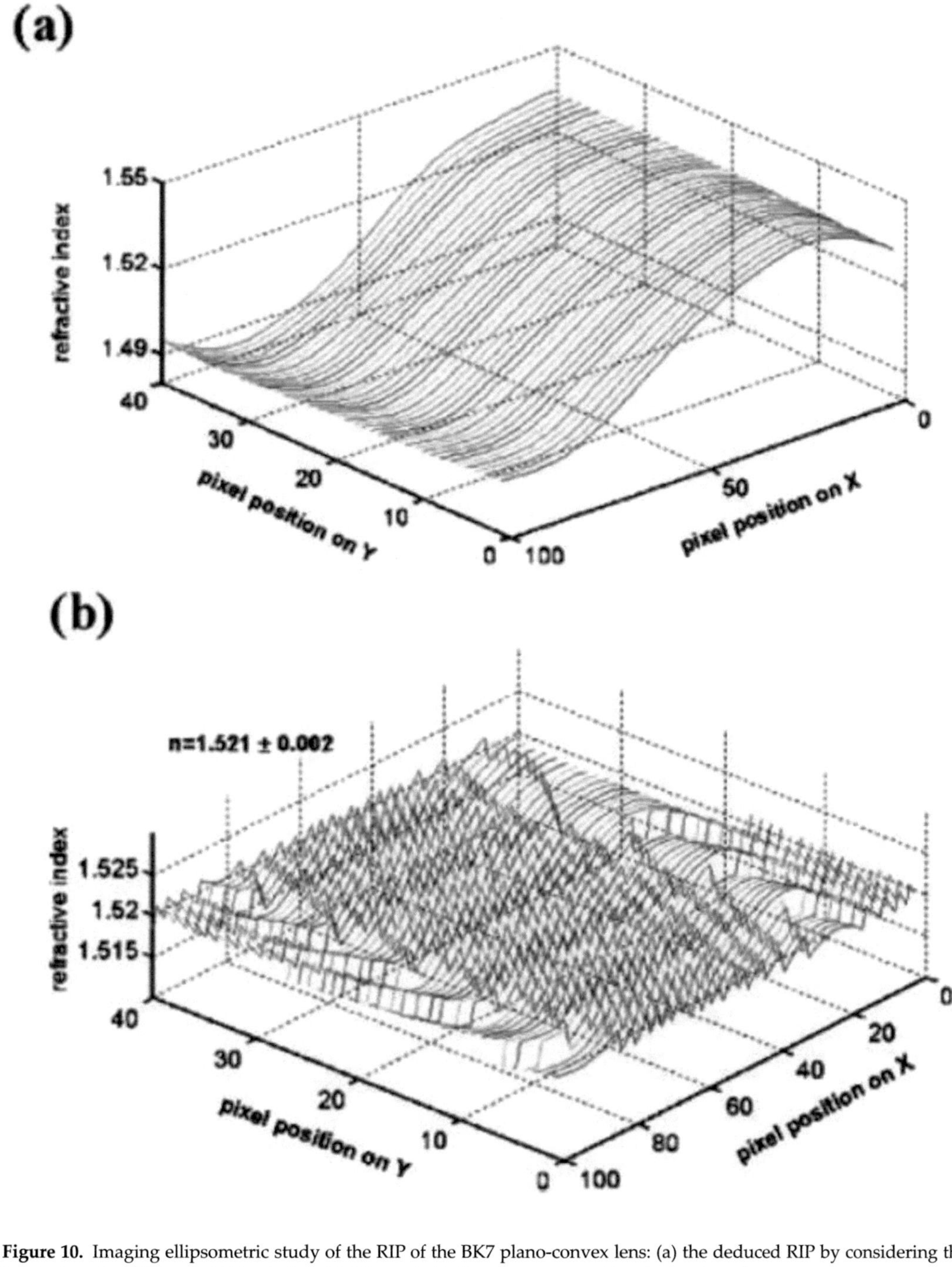


**Figure 10.** Imaging ellipsometric study of the RIP of the BK7 plano-convex lens: (a) the deduced RIP by considering the incident angle at 70° and (b) the deduced RIP by substituting its corresponding incident angles.

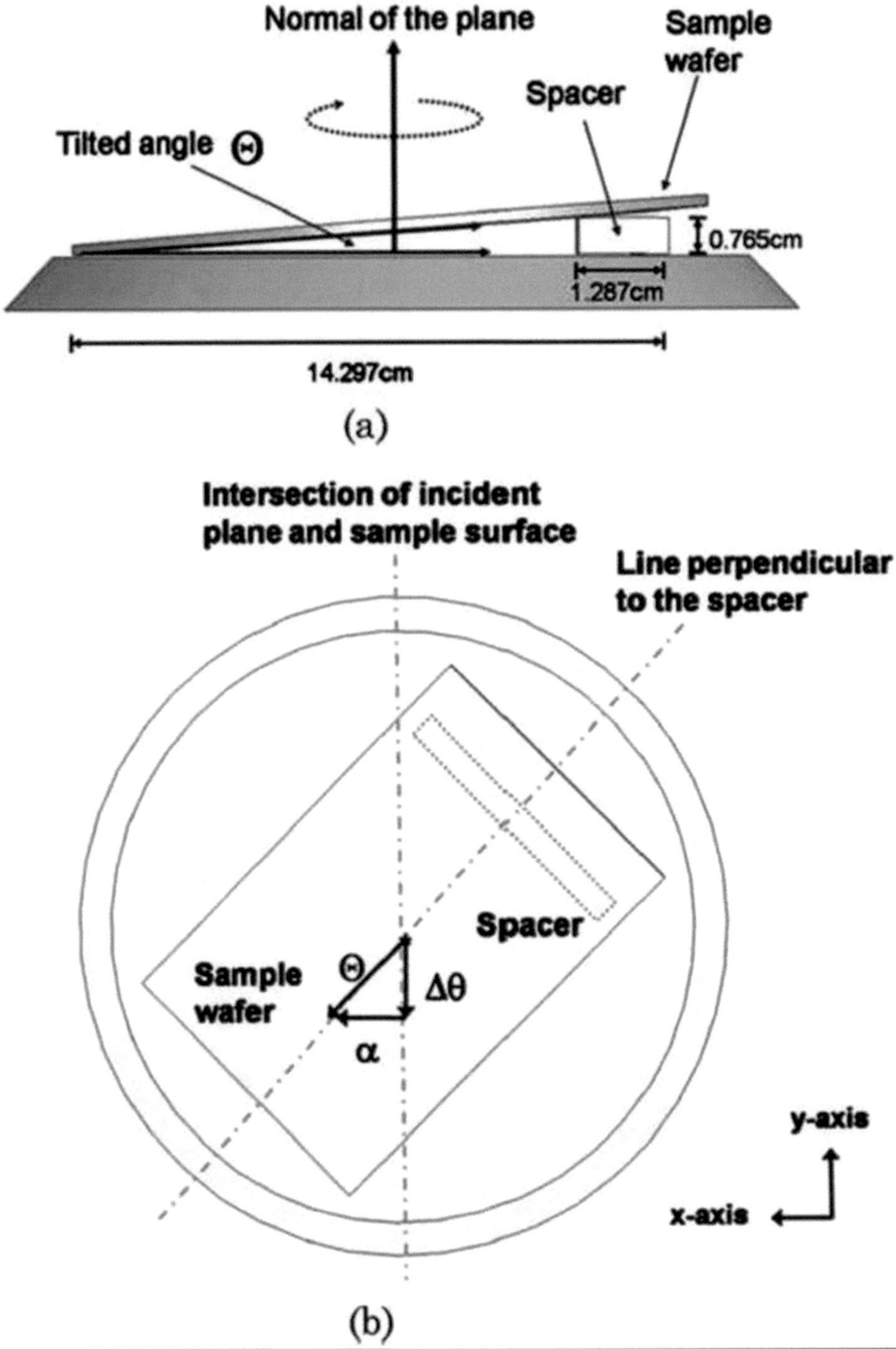


**Figure 11.** Schematic configuration of a tilted sample: (a) side view and (b) top view that indicates the motion of its surface normal: Θ, the tilted angle by spacer; Δθ, the component that contributes to the incident angle variation; α, the component measured in the deviation of the polarizer.

| Condition | Ψ (deg) | Δ (deg) | α (deg) | Δα (deg) | θ[a] (deg) | Δθ (deg) | Θ[a] (deg) |
|---|---|---|---|---|---|---|---|
| Untilted | 69.28 | 90.35 | 0.02 | 2.59 | 69.76 | 1.35 | 2.94 |
| Tilted | 69.68 | 99.42 | −2:57 | | 68.41 | | |
| Untilted | 69.08 | 90.40 | 0.06 | 2.75 | 69.73 | 1.06 | 2.94 |
| Tilted | 66.69 | 97.71 | −2:69 | | 68.67 | | |
| Untilted | 69.24 | 90.42 | 0.06 | 2.95 | 69.75 | 0.08 | 2.94 |
| Tilted | 68.89 | 89.71 | −2:89 | | 69.83 | | |

[a]θ, the incident angle; Θ, the tilted angle; the relationship of the three-direction cosine is $\cos^2\Theta = 1-[\sin^2\Delta\alpha + \sin^2\Delta\theta]$.

**Table 3.** Ellipsometric measurements of tilted and untilted samples.

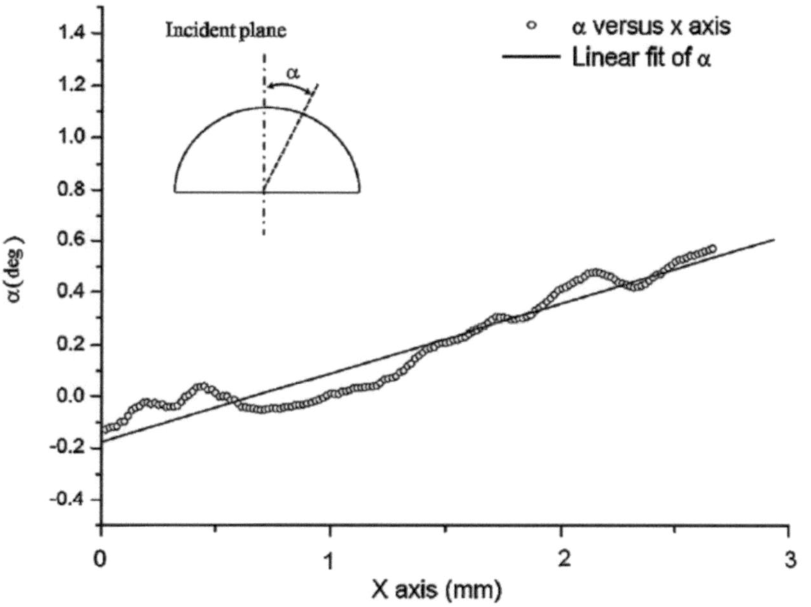


**Figure 12.** Deviation angle of the surface normal across the vertex of a cylindrical lens: open circles, measured α across the vertex of the lens; solid line, its best linear fit.

A cylindrical lens is used to produce a fan-shaped beam for multiple incident angles configuration. This can simplify the analysis to be a two-dimensional case. The three-intensity measurement technique can measure the ellipsometric parameters against each incidence angle but do not have to calibrate the azimuth errors of polarizer and analyzer. As a result of multiple incident angles approach, the whole Ψ curve around the SPR region can be obtained without rotating the sensor chip. The basic setup of the SPR sensor chip is in Kretchmann's configuration, such as shown in **Figure 14**. The beam can be expanded by a beam expander, then focused it on the base of a prism/biosensor around the resonance incident angle (air: 44°). One can measure the three intensities when P is at 45° and −45°, respectively. The measured intensities

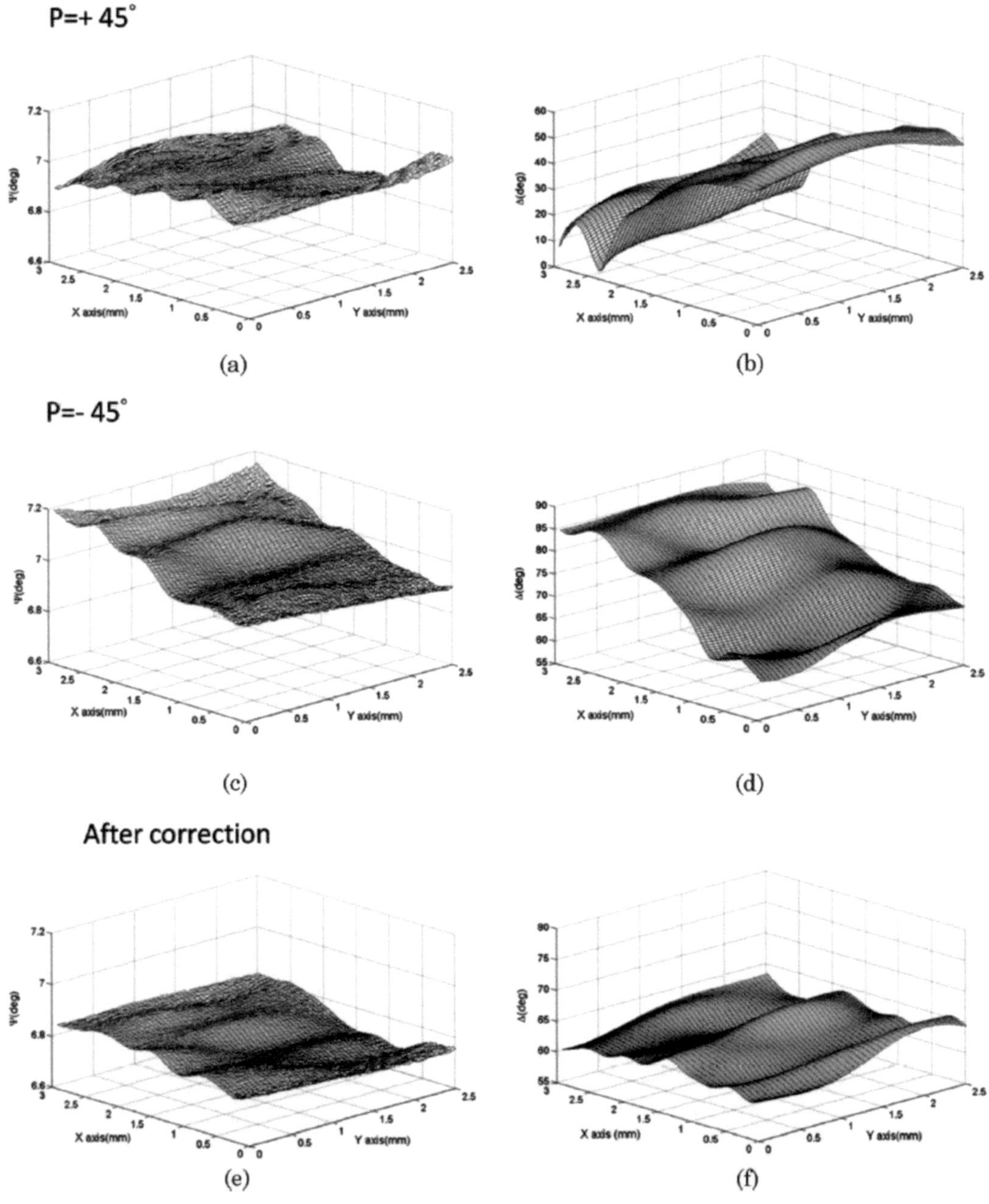


**Figure 13.** Ellipsometric parameters and thickness profile of the cylindrical lens: (a) Ψ and (b) Δ measured when P = 45°; (c) Ψ and (d) Δ measured when P = −45°; (e) Ψ and (f) Δ after correction. Film thickness is 100±4.3 nm.

can be converted into ellipsometric parameters Ψ and Δ. Since the SPR phenomenon is more clearly observable by Ψ, we would like to present it by the distribution of Ψ versus the incident angle. The sensor chip is in Au/air (**Figure 15**) assembly. The solid line in the figure represents the theoretical value. This result can be used for calibration before other measurements.

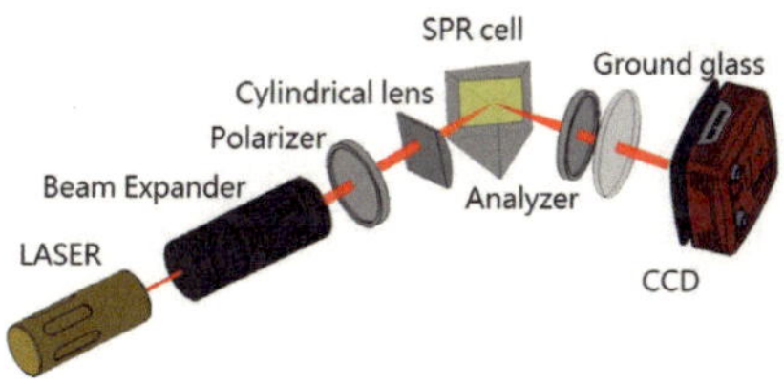


**Figure 14.** The schematic setup of SPR for imaging ellipsometry.

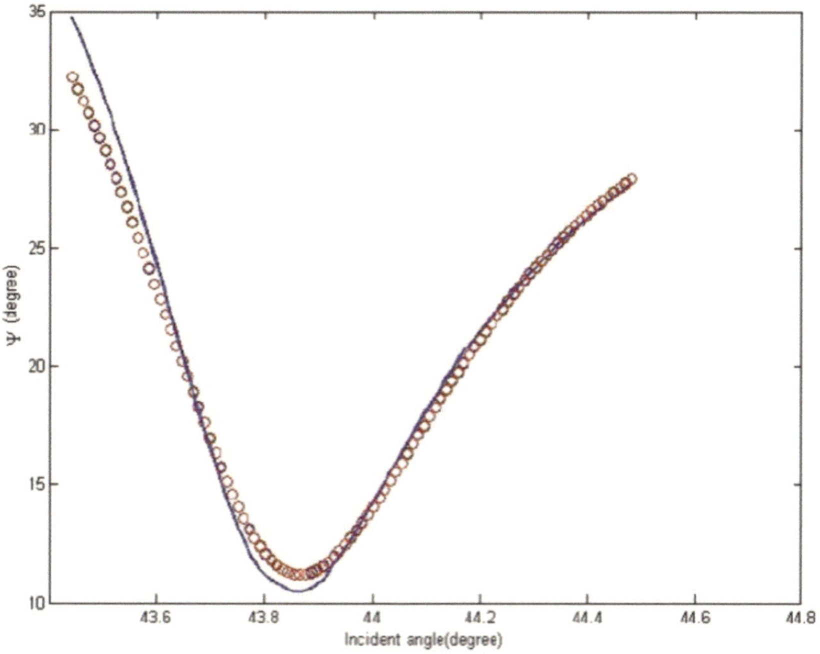


**Figure 15.** The distribution of Ψ versus incident angle of the sensor chip in air (circles are the measured value; solid line is the theoretical value).

## 4. PEM ellipsometry

In PSA ellipsometry, one has to rotate both polarizer and analyzer in order to measure the ellipsometric parameters. These rotations do cause the incident angle to deviate from the setup and slow down the speed of measurements. In the beginning of 1990, the real-time/in situ ellipsometric system has started to develop. By inserting a PEM in the PSA ellipsometry, one can construct an ellipsometry without any mechanical motion and thus substantially increase the speed of measurement. A photoelastic modulator (PEM) is a device that utilizes the photoelastic effect to modulate the phase retardation in a harmonic form [14]. For accurately measuring the ellipsometric parameters with high precision, one has to align the strain axis of PEM to the incident plane and calibrate its modulation amplitude. Since the incident plane has been well aligned in the PSA ellipsometry, one can easily align the strain axis to the incident plane by null and then determine its axis directly through the measurements of two DC intensities when the azimuths of the analyzer are separated by 45°. Regarding the modulation amplitude of PEM, although one can use the waveform of its half-wave in an oscilloscope to calibrate the modulation amplitude, it is usually very hard to visualize the flat shape of the waveform while its modulation amplitude is at half-wave. Recently, there is great progress in the data acquisition (DAQ) system with multi-functions. In this chapter, we perform the two options of the DAQ system for aligning the strain axis and calibrating the modulation amplitude of PEM in the ellipsometry: (1) lock-in amplifiers and (2) digitized oscilloscope. Since the digitized waveform can be stored for analyzing, the system errors of PEM controller can be inspected.

### 4.1. Alignment of the strain axis of PEM

The basic setup of the PEM ellipsometer is shown in **Figure 16**. If the azimuth angle of the initial linear polarized light is set at 45°, and the strain axis of PEM deviates from the incident plane by θ, the measured intensity can be simplified as

$$I(A) = I_0\left[L\sin^2 A + M\cos^2 A\ \tan^2\Psi + N\sin A\ \cos A\right] \quad (19)$$

where

$$L = 0.5[1 + \cos\Delta_P + (1 - \cos\Delta_P)(1 - \sin\Delta_P)] \quad (20)$$

$$M = 0.5[1 + \cos\Delta_P + (1 - \cos\Delta_P)(1 + \sin 4\theta)] \quad (21)$$

$$N = 0.5[1 + \cos\Delta_P - (1 - \cos\Delta_P)\cos 4\theta]\tan\Psi\ \cos\Delta - \sin\Delta_P\ \cos 2\theta\ \tan\Psi\ \sin\Delta \quad (22)$$

The modulation amplitude $\Delta_p$ of PEM is modulated as $\delta_0\cos(\omega t)$. By setting the azimuth angle of analyzer at 0°, the measured intensity can be expressed as

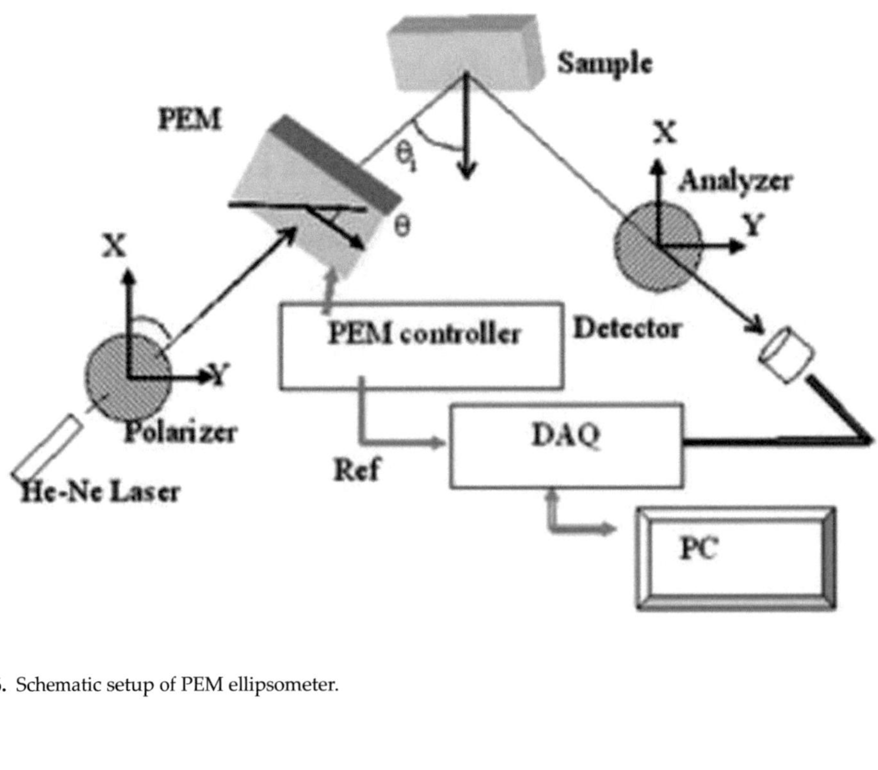


**Figure 16.** Schematic setup of PEM ellipsometer.

$$I(0°) = 0.5I_0 \tan^2\Psi \ [2 + (1 - \cos\Delta_P)\sin 4\theta] \quad (23)$$

then the intensity can be Fourier expanded by its harmonic function

$$\cos\Delta_P = J_0(\delta_0) - 2J_2(\delta_2)\cos 2\omega t\ldots \quad (24)$$

By taking the zero-order Bessel function $J(\delta_0)$ at its zero point, that is, $\delta_0$ =0.383λ, we can simplify the DC component of its intensity as

$$I_{dc}(0°) = 0.5I_0 \tan^2\Psi \ [2 + \sin 4\theta] \quad (25)$$

From this equation, the strain axis of PEM deviates from zero by θ°, which can be measured by the DC component of the intensity. Taking two $I_{dc}$ at $\theta_0$ and $\theta_0$+45°, one can obtain $\theta_0$ through the following relation:

$$\sin 4\theta_0 = 2\frac{I_{dc}(0°)_{\theta=\theta_0} - I_{dc}(0°)_{\theta=\theta_0+45°}}{I_{dc}(0°)_{\theta=\theta_0} + I_{dc}(0°)_{\theta=\theta_0+45°}} \quad (26)$$

A typical result after rough alignment is shown in **Figure 17**. According to the intensity ratio of Eq. (26), one can easily prove that the deviation of azimuth $\delta\theta_o$ resulting from those intensity fluctuations is

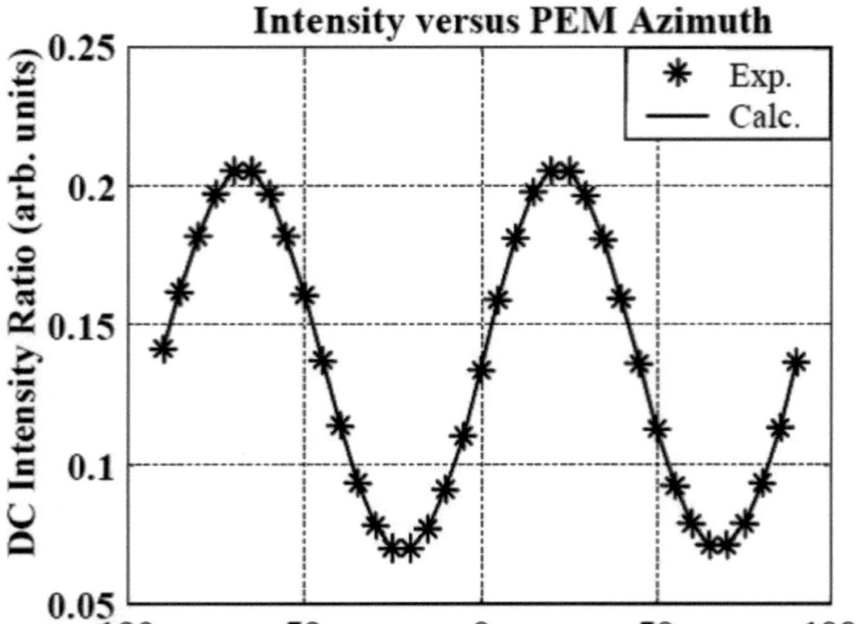


**Figure 17.** DC intensity distribution of the PEM ellipsometer. The azimuth angle of A is at 0° versus the azimuthal angle of the strain axis of PEM. *The experimental data and the solid line show the intensity distribution when the strain axis of PEM is at 0.2764±0.0003°.

$$\delta\theta_0 = \frac{\tan 4\theta_0}{4\tan^2\Psi}\frac{\delta I}{I} \tag{27}$$

The intensity fluctuation can be large at $\theta_0 = 22.5°$, which is far away from the roughly aligned value. By readjusting the $\theta_0$ to zero, one can obtain a well-aligned PEM ellipsometry.

### 4.2. Calibration of the modulation amplitude of PEM

According to the last section, the strain axis of PEM can be well aligned to the incident plane, θ is zero, and the azimuth of polarizer is at 45°, Eq. (19), can be simplified as

$$I(A) = 0.5I_0\left[\sin^2 A + \tan^2\Psi\ \cos^2 A - \tan\Psi\ \sin 2A\ (\cos\Delta\ \cos\Delta_P + \sin\Delta\ \sin\Delta_P)\right] \tag{28}$$

Since the Fourier expansions of the harmonic functions are

$$\sin\Delta_P = 2J_1(\delta_0)\sin\omega t + 2J_3(\delta_0)\sin 3\omega t + \ldots \tag{29}$$

$$\cos\Delta_P = 2J_0(\delta_0) + 2J_2(\delta_0)\sin 2\omega t + 2J_4(\delta_0)\cos 4\omega t + \ldots \tag{30}$$

One can prove the following relation for the measurement of Ψ and Δ:

$$\sin 2\Psi = \frac{\sqrt{\left(\frac{I_{1f}(45°)}{J_1(\delta_0)}\right)^2 + \left(\frac{I_{2f}(45°)}{J_2(\delta_0)}\right)^2}}{I_{dc}(45°) + I_{dc}(-45°)} \tag{31}$$

$$\tan\Delta = \frac{I_{1f}(45°)J_2(\delta_0)}{I_{2f}(45°)J_1(\delta_0)} \tag{32}$$

It is very interesting to notice that the even/odd harmonics are related by the similar physical parameters except the orders of its Bessel function. So,

$$\frac{I_{1f}}{I_{3f}} = \frac{J_1(\delta_0)}{J_3(\delta_0)} \tag{33}$$

$$\frac{I_{2f}}{I_{4f}} = \frac{J_2(\delta_0)}{J_4(\delta_0)} \tag{34}$$

The modulation amplitude $\delta_0$ can be measured by the intensity ratio of even/odd harmonics. In the operation menu of PEM [15], the value $\delta_0$ can be calibrated by the oscilloscope waveform of a half-wave, that is, $\delta_0 = 0.5$, such as shown in **Figure 18**. However, when one is constructing a new element in a system, some defects are expected. Since one can measure the modulation amplitude by the even/odd intensity ratios, one can calibrate the modulation amplitude and compare it with the digitized oscilloscope waveform obtained from DAQ system, such as shown in **Figure 19**. The system errors, azimuth angle, and modulation amplitude can be corrected after the alignment and calibration in order to establish an accurate PEM ellipsometer.

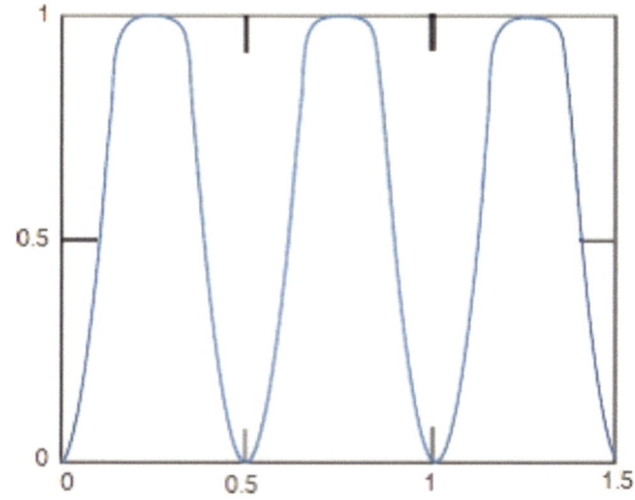


**Figure 18.** The oscilloscope waveforms when the modulation amplitude of PEM is a half-wave [15].

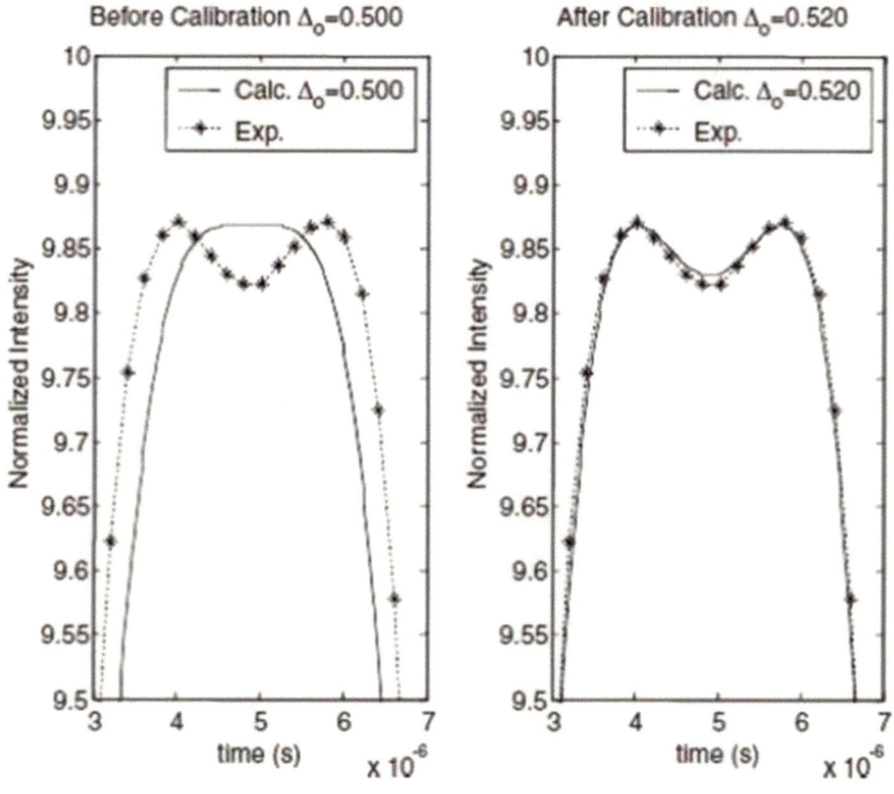


**Figure 19.** Recorded digitized oscilloscope waveform when the modulation amplitude of PEM has been claimed as half-wave: left—before calibration and right—after calibration.

### 4.3. Applications

For monitoring the physical processes, plasma etching [16], deposition [17], thermal heating [18], and photo-induced [19], one has to establish the in situ alignment technique for PEM ellipsometry. A DAQ card is an analog-to-digital converter. In the PEM ellipsometry, the multi-function DAQ (typical PCI-6111) can be used as two lock-in amplifiers for real-time measurement (10 sets/s), and post-flight analysis, that is, analyze the reflected waveform of the digital oscilloscope specification of the DAQ system, so the data rate can reach 25,000 sets/s. With the help of a stroboscope to modulate the light source, one can lock the wave at four specific phases to measure the ellipsometric parameters of the testing sample. By using CCD as the detector, an imaging ellipsometry can also be established in the PEM ellipsometry. The followings are examples for the applications of PEM ellipsometry.

#### *4.3.1. Plasma etching*

For monitoring the etching process, a plasma-etching chamber can install a PEM ellipsometer to monitor the etching process, such as the schematic setup shown in **Figure 20**. A thick wafer is used to extract the etched thickness under etching, which can be monitored by the PEM ellipsometry to avoid over-etch. After calibrating the modulation amplitude, one can etch the film to the targeted thickness by monitoring the behavior of Ψ and Δ with the theoretical predication in real time. For example, we target a thick $SiO_2$ on polysilicon started from around 500 nm to be etched to 30 nm. The measured value with the theoretical value, such as

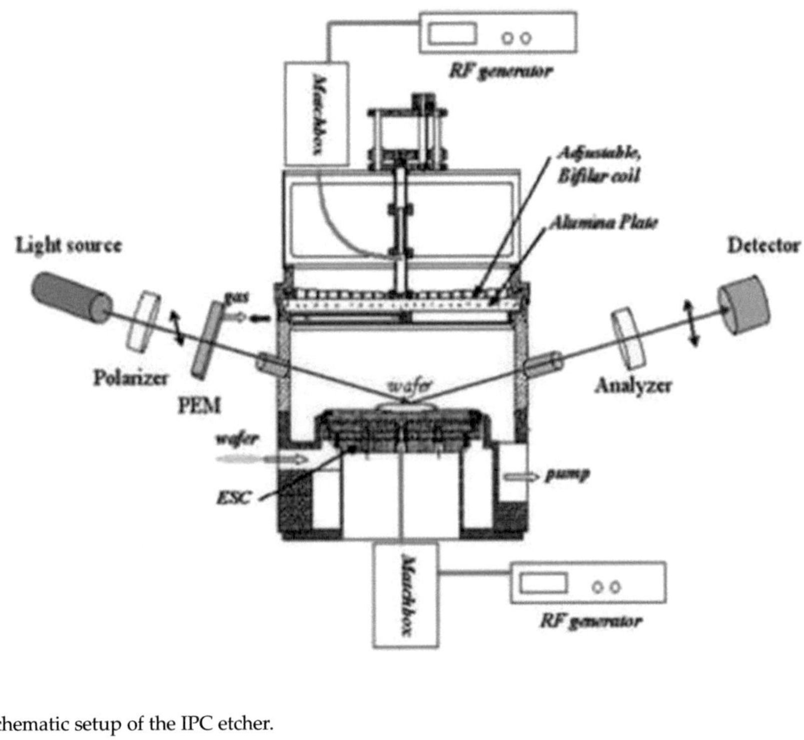


**Figure 20.** Schematic setup of the IPC etcher.

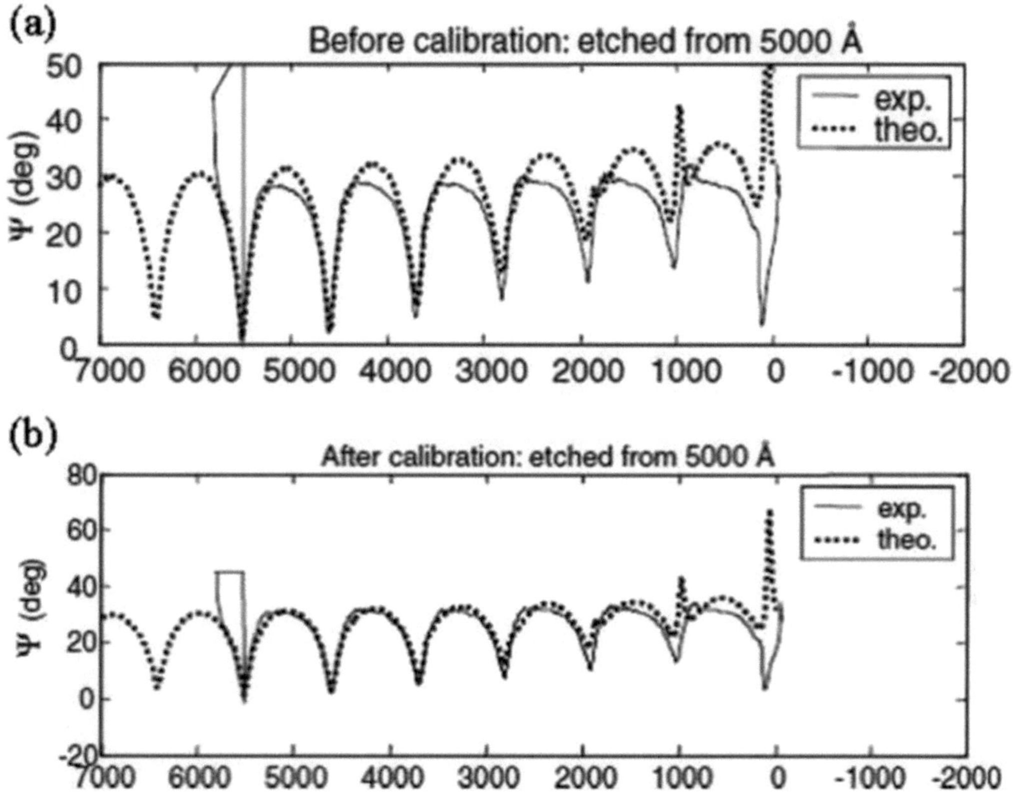


**Figure 21.** Distributions of Ψ for film etched from 500-nm thickness of $SiO_2$ on silicon substrate (a) before calibration and (b) after calibration: (dotted line) theoretical value; (solid line) measured value × fc [16].

in **Figure 21a**, the histogram of Ψ before correction, and **Figure 21b**, the histogram of Ψ after correction (fc, the correction factor from the measurement of modulation amplitude), can be compared. In the corrected curve, we observed the behavior of etching: the sharp rising of Ψ in the end cycle, around 70 nm. In the end, we can control the etching process: the etching stopped at the thickness of the film to be close to 30 nm within 0.2 nm, such as shown in **Figure 22**.

### *4.3.2. Thermal heating*

The perovskite materials are attractive because they exhibit extremely high piezoelectric coefficients and a wide region of controlled dielectric constants when the compositions are near the morphotropic phase boundary (MPB) [20]. In previous studies, doping Ru into complex oxides can enhance the photorefractive effect in the red and near-infrared spectral regions [21]. In addition, using Ru as a dopant in various inorganic crystals can considerably improve their response time and photoconductivity [22]. Therefore, Ru-doped perovskites can be considered as practical material for optical memories. In the process of storage, the temperature of the surroundings can fluctuate due to laser irradiation, heat dissipation, and so on. However, for

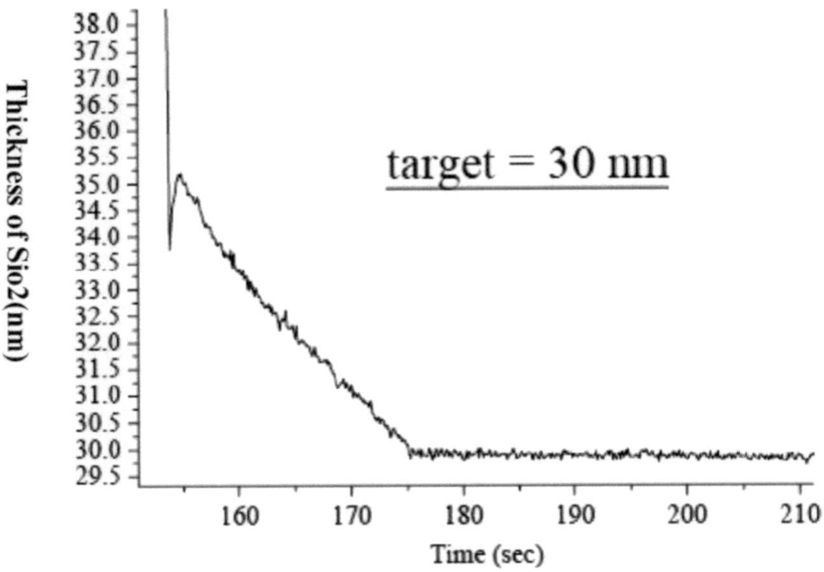


**Figure 22.** The histogram of an etching of $SiO_2/Si$, the target of the process was 30 nm.

optical storage, the refractive index of the perovskite material must be stable under the working temperature. It will be interesting to measure their refractive indices during heating in order to understand what would be the most favorable working temperature for optical storage. One can setup the PEM ellipsometry as shown in **Figure 23** to monitor the refractive index of perovskite material under heating. In the thermally isolated chamber, the samples can be heated from room temperature to 200°C by the thermoelectric cooler (TE cooler), which operates based on the Peltier effect, so it can be used as heater when the electric current is

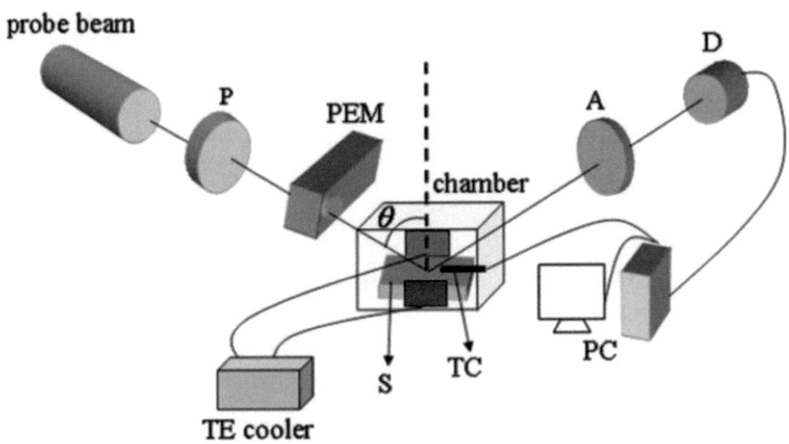


**Figure 23.** Schematic setup of the PEM ellipsometer for measuring the variation in the refractive index during the heating process: probe beam: HeNe laser (632.8 nm); P: polarizer; PEM: photoelastic modulator; S: sample; θ: incident angle.

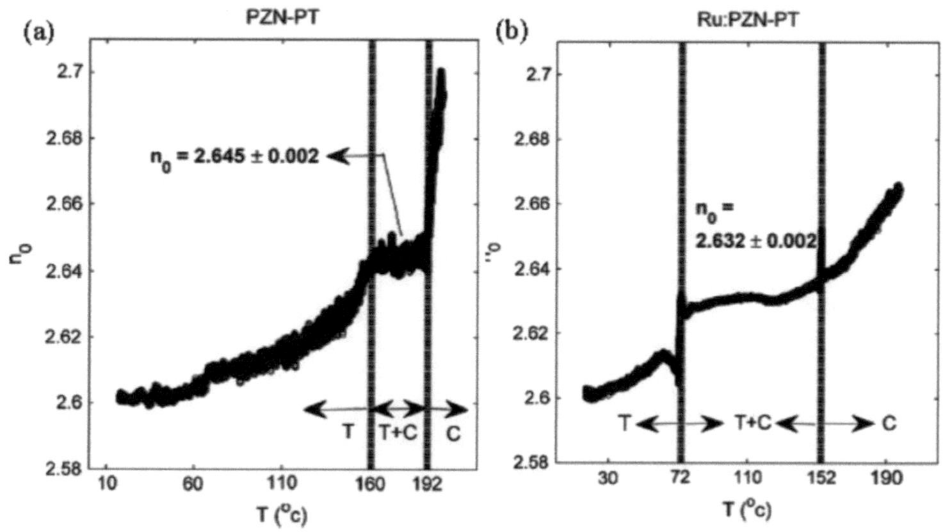


**Figure 24.** Refractive indices of (a) pure 0.9PZN-0.1PT and (b) Ru-doped 0.9PZN-0.1PT against temperature during heating: T: tetragonal; C: cubic.

inversely applied. The TE cooler is small but highly reliable. The sample can be sandwiched between two TE coolers to maintain uniformity of temperature of the sample. A k-type thermocouple was attached to the surface of the sample to monitor the temperature. The voltage produced by the thermocouple can be acquired by one of the channels of the same data acquisition system. The voltage is then converted to the required temperature scale according to the NIST ITS-90 Thermocouple Database. In this way, both the temperature and the ellipsometric parameters can be simultaneously monitored. According to the study by Chuang et al. [18], the ellipsometric parameters of pure and Ru-doped 0.9PZN–0.1PT were measured by PEM ellipsometry with a sampling rate of 10 sets/s. The refractive indices were converted from the measured $\Psi$ and $\Delta$, such as shown in **Figures 24a** and **b**. From the results of the experiment, one can conclude that when Ru is used as a dopant in 0.9PZN–0.1PT, the Curie region is broadened and shifted closer to room temperature. Therefore, the PEM ellipsometry can be used to analyze the behavior of material under heating.

### 4.3.3. *Photo-induced effect*

Holographic data storage has been considered to be a promising data storage technology because it provides high storage density and fast readout rate. One of the fundamental issues for this technique to be successful is the availability of thick recording materials. Recently, research involving organic photopolymers has become of interest because of the flexibility for fabrication. Photopolymers exhibit a photorefractive effect, and the photo-induced changes in the refractive index can be used to record the holographic interference pattern in order to form phase grating. Based on this idea, many new photopolymers have been synthesized and studied. The photo-induced refractive index variation and photo-induced birefringence are two of the most often used mechanisms for producing the necessary changes in refractive

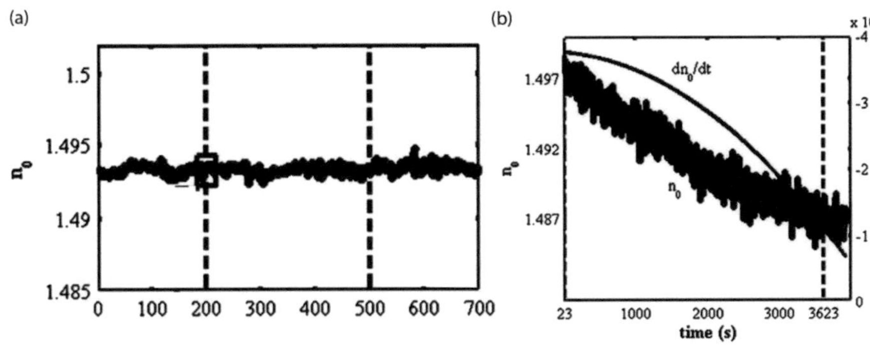


**Figure 25.** Histogram of measured refractive index during heating: (a) pure PMMA: the 488-nm laser was turned on at the 200th second and then turned off at the 500th second. The average refractive index is 1.493±0.003; (b) PQ-doped PMMA: the 488-nm laser was turned on at the 23rd second and turned off at the 3623rd second; the refractive changes from 1.497 to 1.487. It is exposed to 488 nm (Ar/Kr tunable laser) with total exposure energy density of 68.4 J/cm$^2$.

indices for holographic recording. Thus, various methods have been proposed for investigating these two effects in different photopolymers for further improving the properties of the recording materials. The refractive indices of the photopolymers have been measured by an Abbe refractometer and a prism coupler, but the variation of the refractive index is commonly extracted from the diffraction formula by measuring the diffraction efficiency of a recorded grating [23]. Since the PEM ellipsometry can measure the refractive index during exposure, one can compare the photorefractive effect of photopolymers under irradiation. The PEM ellipsometry was used to measure the pure and doped PMMA, such as shown in **Figures 25a** and **b**. From these two measurements, one can conclude that the holographic recording in PQ-doped PMMA was mainly due to the change in the refractive index. The change in the refractive index is produced by molecular structure changes of the PQ molecules. This in situ/real time can observe the structure change under irradiation.

### *4.3.4. Post-flight analysis*

Since most data acquisition systems are multi-function systems, if the PEM ellipsometry employs its function of lock-in amplifier to measure dc, 1f, and 2f signals for deducing the ellipsometric parameters, one can only obtain a data rate of 10 sets/s. In Section 4.2, the DAQ system has been used as digital oscilloscope for calibrating the modulation amplitude of PEM. The waveform obtained from the digital oscilloscope can be stored for post-flight analysis. The ellipsometric parameters can be obtained by fast Fourier transform of the stored waveform. In this case, the PEM ellipsometry can reach the speed limit of PEM, since Hinds PEM90 operated at an oscillating frequency around 50 kHz, so its data rate can reach 25,000 sets/s, for example, if one wishes to measure the response time of twisted nematic liquid crystals. It is known that the response time of twisted nematic liquid crystals (TN-LC) is in the millisecond regime, the transmitted PEM-DAQ ellipsometry can be used to analyze the variation of its ellipsometric parameters under a square wave. **Figure 26** shows the histogram of the ellipsometric parameters of a TN-LC cell applying a 5-Hz square wave with a step voltage of 5 V. From the graph, one can measure the rising time and fall time of the TN-LC, which are 7.5 and 19 ms, respectively.

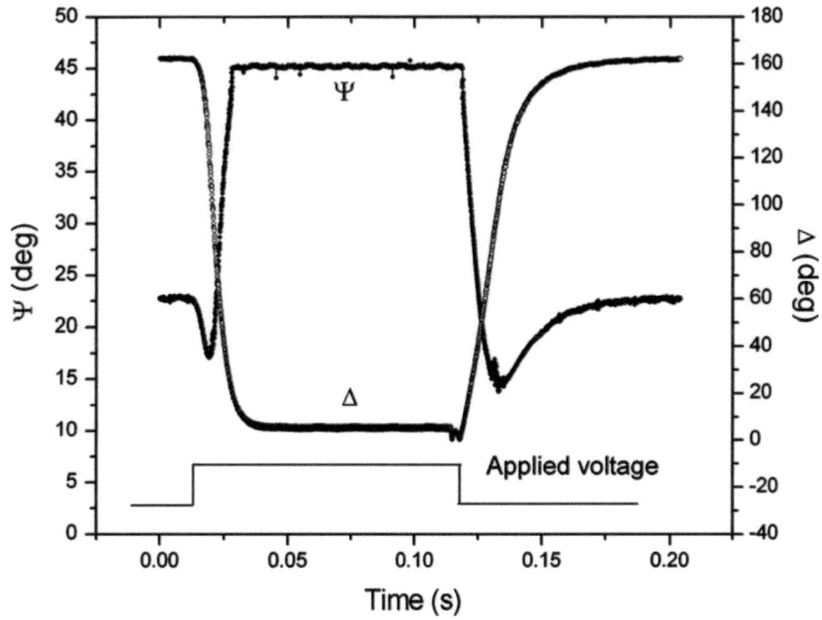


**Figure 26.** Histograms of measured ellipsometric parameters for the twisted nematic liquid crystal cell (transmission ellipsometry, λ = 632.8 nm). A square wave of 5 Hz and strength 5 V was applied.

The generalized ellipsometry (GE) has been developed to measure the refractive indices of anisotropic crystals [24]. In GE, at least three polarization states have to be used for extracting the physical parameters of anisotropic crystals. Using a fixed polarizer and PEM, one can generate various polarization states (200/cycle, Hinds90) because of the harmonic oscillation property of PEM. Two modulators of generalized ellipsometry [25] have been developed to measure the anisotropic medium. Due to the development of inverse problem in optics, Chuang et al. [26] implemented the program of Hodgkinson (BTF tool box) to simulate the reflected intensity profile. This profile can be analyzed and compared with the digitized waveform obtained from DAQ system of PEM ellipsometer. In his study, he found a distinguished characteristic in the reflected waveform for different complex refractive indices. Instead of using the Fourier transform on the waveform, he analyzed the reflected/transmitted waveform to extract the physical parameters. In the same reference, he proved that two incident angles are required to extract the complex refractive indices and the orientation of its optical axis of the anisotropic crystal. First, he concentrated his analysis on the reflected waveform at the incident angle of 70°. According to the Fresnel reflection coefficient of a uniaxial crystal, one can graph the normalized reflected intensity waveform as **Figure 27a**. The asymmetric distribution provides the information of absorbing or nonadsorbing materials. Assume $k_{ave} = 0.5(k_e+k_o)$ and $\Delta k = k_e-k_o$ are the averaged and difference between the extinction of ordinary and extraordinary coefficients, respectively, and $n_{ave} = 0.5(n_e+n_o)$ and $\Delta n = n_e-n_o$ are the averaged and difference between ordinary and extraordinary refractive indices, respectively. Using **Figures 27b** and **c**, one can conclude that the value of averaged extinction coefficient and difference can be estimated from the height (H) of the dip by the parametric

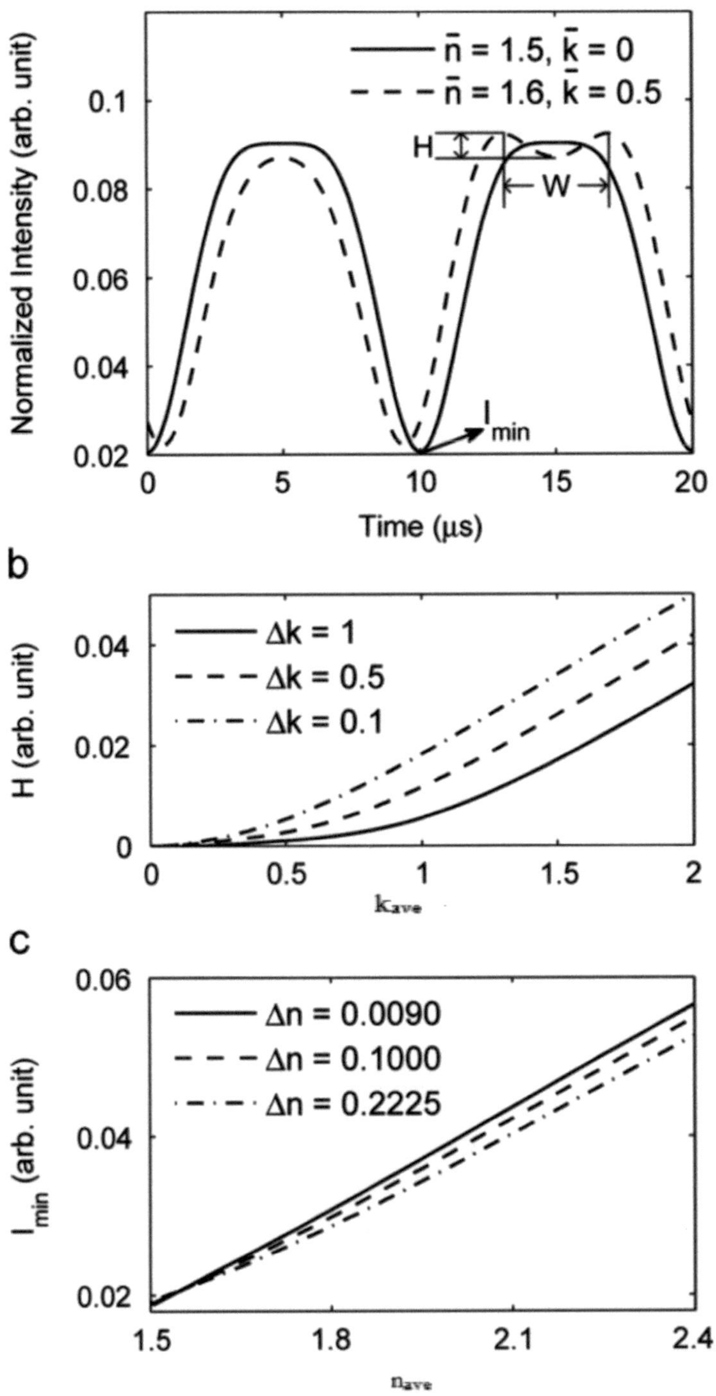


**Figure 27.** Estimation for complex refractive index ($\theta_i$=70°): (a) the normalized reflected intensity profile; (b) the parametric equations of the dip depth (H) versus the averaged extinction coefficient ($k_{ave}$); and (c) the parametric equations of the minimum intensity ($I_{min}$) versus the averaged refractive indices ($n_{ave}$).

| | |
|---|---|
| $k_{ave} = 0.5(k_e+k_o)$ | $H_{70} = p_0 + p_1 k_{ave} + p_2 k_{ave}^2$ |
| $\Delta k = k_e - k_o$ | $p_0 = -0.002 + 0.003\Delta k$ |
| | $p_1 = -0.015 + 0.021\Delta k$ |
| | $p_2 = 0.006 + 0.005\Delta k$ |

**Table 4.** The parametric equations of dip height H for estimating the extinction coefficients obtained by simulated reflected waveform when $\theta_\iota$ at 70°.

| | |
|---|---|
| $n_{ave} = 0.5(n_e+n_0)$ | $I_{70min} = q_0 + q_1 n_{ave}$ |
| $\Delta n = n_e - n_0$ | $q_0 = -0.046 + 0.031\Delta n$ |
| | $q_1 = 0.043 + 0.022\Delta n$ |

**Table 5.** The parametric equations of $I_{min}$ for estimating the refractive coefficient obtained by simulated reflected waveform when $\theta_\iota$ at 70°.

equations (**Table 4**). The averaged refractive index can be estimated by the minimum intensity $I_{min}$ of the waveform, which is linearly related to the average refractive index (**Table 5**). It is known that the genetic algorithm is the mechanism of nature evolution and is considered as the optimal solution in the complex multidimensional search spaces. The parametric equations really can narrow down the searching region in genetic algorithm. Since the genetic algorithm [27] has been applied in ellipsometry, one can simulate the waveform and find the characteristic property to narrow down the searching region, then to search the complex refractive indices and the orientation of optical axis. Using two optimized incident angles, Chuang et al. [26] extracted the $n_o$ and $n_e$ of quartz crystal to be 1.543 and 1.552, respectively. In addition, the orientation of its optical axes was also measured from the same

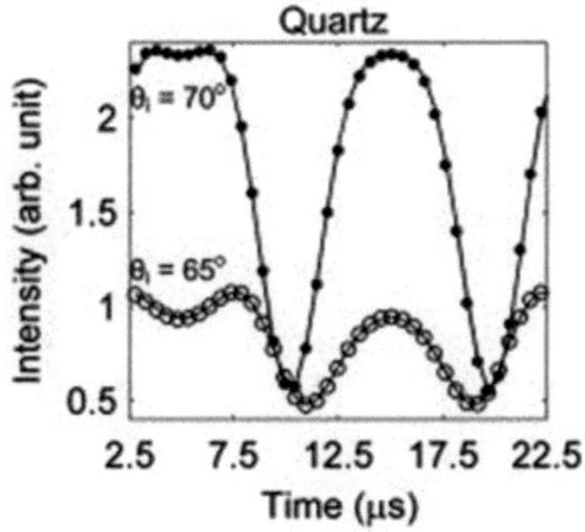


**Figure 28.** Measured reflected intensity profiles of quartz at incident angles of 65° (o) and 70° (•). All these measured values are compared with the calculated values (-). Only 1/5 data points in each cycle are displayed.

waveform. **Figure 28** shows the comparison between simulated waveform by using the extracted physical parameters (refractive indexes, orientation of optical axis) and measured values. This work presents the powerfulness of the post-flight analysis technique; one only has to use one cycle (20 μs of Hinds PEM 90) of the reflected intensity signals to extract all the optical parameters.

## 5. Future prospects

The three-intensity measurements technique in PSA ellipsometry can be easily constructed into the imaging ellipsometry. In this system, the rotation of polarizer and analyzer not only slows down the measuring speed but also causes the incident beam to deviate from its setting. This is the reason for establishing the PEM ellipsometry. However, the modulation frequency of PEM is very high (most PEM operates in 50 kHz), and one must use fast response detector for PEM ellipsometry. The operation of CCD camera is a time-consuming process in comparing with the modulation frequency of PEM. The PEM imaging ellipsometry can only be established by employing the modulated laser diode as pulse for strobe illumination [28] at four specific phases. In this technique, one has to calibrate the initial phase for synchronizing the four specific phases to perform ellipsometric measurements. Again, with the help of symbolic computer program, Mathematica, Tsai [29] de-associated the initial phase by extra strobe measurement. This PEM stroboscopic imaging ellipsometry can measure a moving sample dynamically, such as the motion of dropping oil on $SiO_2$ film. The speed of measurement can be improved by using high-power light source to reduce the exposure time and high-speed DAQ system to match the data transfer of CCD.

For further developing the ellipsometry, one can use the liquid crystal variable retarder to replace PEM. It is interesting to notice [30] a three-intensity measurements technique in ellipsometry by varying the phase retardation of liquid crystal instead of varying the polarization by rotating any part of the system. In addition to a liquid crystal variable retarder, a quarter wave plate has been introduced to form an ellipsometer without any mechanical motion. The in situ calibration of the liquid crystal variable retarder has to be performed. Just like the three-intensity measurement in PSA ellipsometry, one can install a liquid crystal variable retarder by varying three phases to form an imaging ellipsometry. Polarizer and quarter wave plate can be formed as polarization generator, liquid crystal variable retarder and analyzer can be packed as polarization analyzer. Since all these components are light-weight with low cost, one can construct a compact and portable system for imaging ellipsometric measurements. The only disadvantage of liquid crystal variable retarder is its stability in temperature. For accurate measurement, one has to use a temperature controller. We await the improvement of the liquid crystal variable retarder for the establishment.

## Acknowledgements

All these works were accomplished by my students under the grants of National Science Council in Taiwan.

## Author details

Yu-Faye Chao

Address all correspondence to: yfchao@mail.nctu.edu.tw

Department of Photonics, National Chiao Tung University, Taiwan, ROC

## References

[1] Azzam RMA, Bashara NM. Ellipsometry and Polarized Light. Paperback. North Holland: Elsevier Science Publishers B.V.; 1987

[2] Collins RW. Automatic rotating element ellipsometers: Calibration, operation, and real-time applications. Review of Scientific Instruments. 1990;**61**:2029-2062. DOI: 10.1063/1.1141417

[3] Chao YF, Wei CS, Lee WC, Lin SC, Chao TS. Ellipsometric measurements and its alignment: Using the intensity ratio technique. Japanese Journal of Applied Physics. 1995;**34**:5016-5019. DOI: 10.1143/JJAP.34.5016

[4] Chao YF, Lee WC, Hung CS, Lin JJ. A three-intensity technique for polarizer-sample-analyser photometric ellipsometry and polarimetry. Journal of Physics D: Applied Physics. 1998;**31**:1968-1974. DOI: 10.1088/0022-3727/31/16/005

[5] Steel MR. Method for azimuthal alignment in ellipsometry. Applied Optics. 1971;**10**:2370-2371

[6] Wang MW, Chao YF. Azimuth alignment in photoelastic modulation ellipsometry at a fixed incident angle. Japanese Journal of Applied Physics. 2002;**41**:3981-3986. DOI: 10.1143/JJAP.41.3981

[7] Chao YF, Lee KY, Lin YD. Analytical solutions of the azimuthal deviation of a polarizer and an analyzer by polarizer-sample-analyzer ellipsometry. Applied Optics. 2006;**45**: 3935-3939. DOI: 10.1364/ao.45.003935

[8] Chao YF, Wang MW, Ko ZC. An error evaluation technique for the angle of incidence in a rotating element ellipsometer using a quartz crystal. Journal of Physics D: Applied Physics. 1999;**32**:2246-2249. DOI: 10.1088/0022-3727/32/17/315

[9] Tomlinson WJ. Application of GRIN-rod lenses in optical fiber communication systems. Applied Optics. 1980;**19**:1127-1138. DOI: 10.1364/AO.19.001127

[10] Garratt J, Mills M. Measurement of the roughness of supersmooth surfaces using a stylus instrument. Nanotechnology. 1996;**7**:13-20. DOI: 10.1088/0957-4484/7/1/002

[11] D'Acquisto L, Fratini L, Siddiolo AM. A modified moiré technique for three-dimensional surface topography. Measurement Science and Technology. 2002;**13**:613-622. DOI: 10.1088/0957-0233/13/4/326

[12] Han CY, Lee ZY, Chao YF. Determining thickness of films on a curved substrate by use of ellipsometric measurements. Applied Optics. 2009;**48**:3139-3143. DOI: 10.1364/AO.48.003140

[13] Han CY, Du CY, Chen YR, Chao YF. Developing a phase and intensity measurement technique with multiple incident angles under surface plasmon resonance condition. Proceedings of SPIE. September 2013;**8905**:89052N:1-7. DOI: 10.1117/12.2035029

[14] Jasperson SN, Schnatterly SE. An improved method for high reflectivity ellipsometry based on a new polarization modulation technique. Review of Scientific Instruments. 1969;**40**:761-767. DOI: 10.1063/1.1684062

[15] Oakberg T, Wang B. Light intensity modulation using a PEM. Technology for Polarization Measurement. 2000:1-3 http://www.hindsinstruments.com/wp-content/uploads/Light-Intensity-Modulation.pdf

[16] Wang MW, Chao YF, Leou KC, Tsai FH, Lin TL, Chen SS, et al. Calibrations of phase modulation amplitude of photoelastic modulator. Japanese Journal of Applied Physics. 2004;**43**:827-832. DOI: 10.1143/JJAP.43.827

[17] Collins RW. In situ ellipsometry of thin-film deposition: Implications for amorphous and microcrystalline Si growth. Journal of Vacuum Science & Technology. B. Microelectronics and Nanometer Structure. 1989;**B7**:1155-1164. DOI: 10.1116/1.584566

[18] Chuang CI, Marinova V, Lin SH, Chao YF. Phase-modulated ellipsometry for probing the temperature-induced phase transition in ruthenium-doped lead zinc niobate-lead titanate single crystal. Thin Solid Films. 2011;**519**:2867-2869. DOI: 10.1016/j.tsf.2010.12.085

[19] Chuang CI, Hsiao YN, Lin SH, Chao YF. Real-time measurement of photo-induced effects in 9,10-phenanthrenequinone-doped poly(methyl methacrylate) photopolymer by phase-modulated ellipsometry. Optical Communication. 2010;**283**:3279-3283. DOI: 10.1016/j.optcom.2010.04.036

[20] Liu W, Ren X. Large piezoelectric effect in Pb-free ceramics. Physical Review Letters. 2009;**103**:257602 1-4. DOI: 10.1103/PhysRevLett.103.257602

[21] Lin CH, Huang CY, Chang JY. Increasing the conductivity of photorefractive $BaTiO_3$ single crystals by doping Ru. Applied Surface Science. 2003;**208–209**:340-344. DOI: 10.1016/S0169-4332(02)01393-4

[22] Rakitina L, Gospodinov M, Briat B. Photorefractive and photochromic properties of Ru doped Sr0:61Ba0:39Nb2O6 crystal. Optical Communication. 2002;**213**:373-378. DOI: 10.1016/S0030-4018(02)02124-7

[23] Robinson TG, DeCorby RG, McMullin JN, Haugen CJ, Kasap SO, Tonchev D. Strong Bragg gratings photoinduced by 633-nm illumination in evaporated As2Se3 thin films. Optics Letters. 2003;**28**:459-461. DOI: 10.1364/OL.28.000459

[24] Azzam RMA, Bashara NM. Application of generalized ellipsometry to anisotropic crystals. Journal of the Optical Society of America. 1974;**64**:128-133. DOI: 10.1364/JOSA.64.000128

[25] Jellison GE, Modine F. Two-modulator generalized ellipsometry: Theory. Applied Optics. 1997;**36**:8190-8198. DOI: 10.1364/AO.42.003765

[26] Chuang CI, Lin SH, Chao YF. Rapid and accurate extraction of optical parameters for uniaxial bulk media by phase modulated ellipsometry. Optics and Lasers in Engineering. 2013;**51**:861-866. DOI: 10.1016/j.optlaseng.2013.02.004

[27] Cormier G, Boudreau R. Genetic algorithm for ellipsometric data inversion of absorbing layers. Journal of the Optical Society of America A: Optics, Image Science and Vision. 2000;**17**:129-134. DOI: 10.1364/JOSAA.17.000129

[28] Han CY, Chao YF. Photoelastic modulated imaging ellipsometry by stroboscopic illumination technique. Reviews in Scientific Instruments. 2006;**77**:1-5. DOI: 10.1063/1.2173027

[29] Tsai HM, Chen CW, Tsai TH, Chao YF. Deassociate the initial temporal phase deviation provided by photoelastic modulator for stroboscopic illumination polarization modulated ellipsometry. Reviews in Scientific Instruments. 2011;**82**:5-10. DOI: 10.1063/1.3568745

[30] Shih W, Hsieh M, Chao YF. A compact in-situ ellipsometer using the liquid crystal variable retarder. Proceedings of SPIE. September 5, 2014;**9200**:1-7. DOI: 10.1117/12.2060567

Chapter 4

# Multilayer Polarizer at the Energy of 50–1000 eV

Jingtao Zhu and Mingqi Cui

Additional information is available at the end of the chapter

http://dx.doi.org/10.5772/intechopen.70088

**Abstract**

Accurate evaluation of polarization states of the radiation is necessary for polarization-sensitive studies, which requires polarization optical elements, such as polarizer, analyzer, and phase retarder. In extreme ultraviolet (EUV) and soft X-ray region, the closeness of the real part of the refractive index to unity, coupled with high absorption, makes the realization of polarizers such as birefringence and dichroic polarizers impossible. Periodical multilayers are commonly used in polarization study working at the quasi-Brewster angle. To expand narrow spectral bandwidths of periodic multilayers, aperiodic and lateral gradual multilayer polarizers including reflective analyzers and transmission phase retarders are utilized. In this chapter, we demonstrate a series of periodic, aperiodic, and lateral gradual broadband multilayer polarizers with the material combinations of Mo/Si, Mo/Y, Mo/$B_4C$, Cr/C, Cr/Sc, Cr/Ti, Cr/V, $WSi_2$/Si, W/$B_4C$, etc. Different multilayer polarizers correspond to different energy ranges, covering 50–1000 eV totally, including "water window" and the L absorption edges of Fe, Co, and Ni. Polarization measurements are performed at BESSY II, Diamond Light Source, National Synchrotron Radiation Laboratory in Hefei and Beijing Synchrotron Radiation Facility. Some of the polarizers we have developed are applied to the polarization measurements of BESSY II UE56/1-PGM and Beamline 3W1B of Beijing Synchrotron Radiation Facility.

**Keywords:** polarization, extreme ultraviolet (EUV), soft X-ray, multilayer thin film, synchrotron radiation beamline

## 1. Introduction

For polarization-sensitive studies, such as circular dichroism spectroscopy, spin-polarized photoelectron spectroscopy, and spectroscopic ellipsometry [1–4], accurate evaluation of the polarization state of the radiation is clearly crucial. Traditionally, in the visible and ultraviolet regions of the spectrum, birefringence and dichroic polarizers are used. But, in the soft X-ray region and extreme ultraviolet regions, collectively known as the EUV, the closeness of the (real part of the) refractive index, coupled with high absorption, makes the realization of these kinds of polarizers impossible. As the one-dimension artificial crystal, multilayer can be utilized as polarizers and phase retarders in EUV and soft X-ray regions for its interference structures, which are based on Bragg diffraction theory [5, 6]. Periodical multilayers are commonly used in polarization study when they work at the quasi-Brewster angle. But the narrow spectral bandwidth of Bragg peak cannot fulfill the need of broadband spectrum applications. In order to expand the narrow spectral bandwidth of the periodic multilayer, the aperiodic multilayer polarizer including reflective analyzer and transmission phase retarder is utilized [7–10]. Complete polarization analysis has been performed by this new developed polarization system [11]. However, with the energy increasing, especially for the "water window" energy range, which is essential to biology researches, the aperiodic multilayer is no longer suitable because the period thicknesses in such energy region are ultra-thin (typically less than 1.0 nm). In this case, a lateral gradual multilayer is utilized and demonstrated with different material combinations. We also design and fabricate different types of multilayer polarizers at the energy range from 50 to 1000 eV covering the L absorption edges of metal magnetic materials Fe, Co, and Ni and also including the important energy range "water window." Measurements of the multilayer polarizers are performed at BESSY II, Diamond Light Source, Beijing Synchrotron Radiation Facility, and National Synchrotron Radiation Laboratory.

## 2. Design

The design of multilayer for polarization analysis requires determining the following parameters: materials combination, incident angle, and layer thickness distribution. The best material combination for multilayer is to form smooth and abrupt interface with low absorption and high optical contrast. The imperfections of interface as well as the absorption will influence the reflectivity of the multilayer. The absorption will be enhanced at a specific energy during the interaction between materials and incident light. To minimize the absorption, we choose material with the absorption edge slightly smaller than the working energy as the spacing layer. To maximize the interference in the structure of multilayer, the one with high optical contrast to the spacing layer is selected as the absorption layer. Here, we design periodical multilayer polarizers with different material combinations at the energy range from 50 to 1000 eV. **Table 1** shows the typical material combinations at different working energies.

Taking Mo/Y multilayer as an example, the calculated s-reflectivity, p-reflectivity, and $R_s/R_p$ with the optimized parameters are shown in **Figure 1**. *Y*-axis is in log to make all curves clear. The corresponding parameters are shown in **Table 2**, where, *D* represents the period thickness, *N* for bilayer numbers, and $\gamma$ for the ratio of spacing layer thickness to period thickness. From **Figure 1**, we can see that the bandwidth is quite narrow. Such polarizers are effective over a very small wavelength range and close to the Brewster angle (approximately 45° for EUV and soft X-ray radiation); the narrowband wavelength and angular properties may be disadvantages for some studies. In order to overcome this, Yanagihara et al. [12] developed the

http://dx.doi.org/10.5772/intechopen.70088

| Energy (eV) | Wavelength (nm) | Material combination |
|---|---|---|
| 50–98 | 24–12 | Mo/Si |
| 98–155 | 8–12 | Mo/Y |
| 155–185 | 6.7–8 | $Mo/B_4C$ |
| 177–282 | 4.4–6.7 | Cr/C |
| 284–543 | 2.3–4.4 | Cr/Sc, Cr/Ti, Cr/V |
| 540–1000 | 1.2–2.3 | $W/B_4C$ |

**Table 1.** Typical material combinations for different energy ranges.

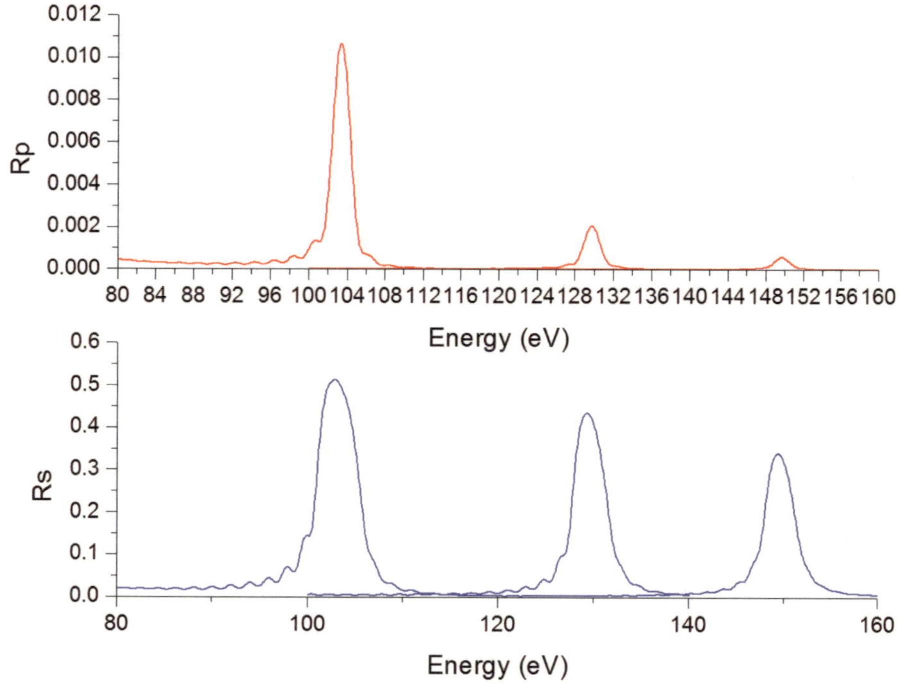


**Figure 1.** $R_s$, $R_p$, and $R_s/R_p$ for Mo/Y periodic multilayers.

| Energy (eV) | Material combination | $N$ | $\gamma$ | $D$ (nm) | $R_s$ |
|---|---|---|---|---|---|
| 100 | Mo/Y | 40 | 0.56 | 9.60 | 0.58 |
| 130 | Mo/Y | 60 | 0.50 | 7.05 | 0.46 |
| 150 | Mo/Y | 80 | 0.53 | 6.03 | 0.34 |

**Table 2.** Parameters of Mo/Y periodic multilayers.

double-multilayer polarizer, which, however, requires a complicated experimental arrangement and degrades the throughput. By rotating and translating the multilayer polarizers, an s-reflectivity of >4% in the wavelength range of 8–15 nm has been achieved. Also, Kortright et al. [13] have constructed a continuously tunable, laterally graded multilayer-based polarimeter working over a wide energy range. Using a series of translatable laterally gradual multilayers, the polarization state of the beam can be determined over the range of 1.6–25 nm.

To simplify the use of broadband polarizers, the design and performances of aperiodic broadband multilayer polarizers are described here.

At the quasi-Brewster angle, the ratio of the s- to p-reflectivity reaches the largest, where the reflective analyzer can be obtained. Since the complex refractive indices of all materials are very close to unity in EUV and soft X-ray ranges, the quasi-Brewster is close to 45°. The polarization degree $P$ of beam reflected from a multilayer is determined by usual manner,

$$P = \frac{R_s - R_p}{R_s + R_p} \tag{1}$$

where $R_s$ and $R_p$ are the reflectivities for s- and p-polarized radiation, respectively. In the design of a reflective multilayer analyzer, the optimization of layer thickness distribution is achieved by minimizing the merit function (MF) [14],

$$\mathrm{MF} = \frac{1}{m}\sum_{j=1}^{m}\left[1 - P(\lambda_j)\right]^2 \tag{2}$$

where the summation is over a selection of discrete wavelengths in the desired range. The layer thickness distribution is considered as independent variable. During the recursive optimization, only randomly selected layer thickness changes that decrease MF are retained, finally leading to an optimized layer thickness distribution that provides a minimum value of MF.

For the wide angular multilayer analyzer, the optimization is achieved by minimizing the following merit function (MF)

$$\mathrm{MF} = \frac{1}{m}\sum_{j=1}^{m}\left(1 - P(\theta_j)\right)^2 \tag{3}$$

where the summation is over a selection of discrete angles in the desired range.

The phases of the transmitted s ($\Phi_s$) and p ($\Phi_p$) electromagnetic fields can be calculated following the formalism of Vidal and Vincent [15], in which the phase shift $\Delta\Phi = \Phi_s - \Phi_p$ is evaluated as a function of the wavelength, the grazing angle, the optical constants of the materials, the number of layers, and their thicknesses. Initially, the published criteria for selecting material and the number of bi-layer were used to satisfy both the maximum phase shift and the transmission intensity [16]. Then, the grazing angle was set in the region between the Bragg peak and the total reflection [17], at which the phase shift is maximum. The numerical optimization method was based on the minimization of the usual merit function (MF)

$$MF = \left(\frac{1}{n}\sum_{j=1}^{n}\left(\Phi_0 - \Delta\Phi(\lambda_j)\right)^2\right)^{1/2} \quad (4)$$

where $\Delta\Phi(\lambda_j)$ is the calculated phase retardation at wavelength $\lambda_j$ and $\Phi_0$ is the desired phase shift.

By using the above scheme, we have designed a series of aperiodic multilayers such as Mo/Y [7], Mo/Be, Ni/C, and Mo/Si [8, 9], which can be utilized as broadband reflection analyzer and transmission phase retarder [10]. The broad angular Mo/Si multilayer analyzer is described in detail as an example, which the layer thicknesses oscillate from 3 to 11 nm, a range that is feasible to manufacture using direct current (DC) magnetron sputtering machine. The calculated mean s-reflectivity in the 45–49° range at working wavelength of 13 nm is 64.99 ± 0.18% and the degree of polarization is up to 99.12 ± 0.71%.

However, with the increasing energy, especially in the "water window" energy range, period thicknesses are decreasing to around 1.0 nm. It is difficult for magnetron sputtering to deposit such thin layer accounting interface roughness and diffusion. In this case, a lateral gradual multilayer structure is utilized to fulfill the Bragg condition either for varying energy or varying incident angle. Each periodic thickness of the lateral gradual multilayer varies in the horizontal direction corresponding to different energy, so that the energy bandwidth can be broadened to a respectively large range.

We design a series of lateral gradual multilayer polarizers with the material combinations of Mo/Si, Mo/Y, Mo/$B_4C$, Cr/C, Cr/Sc, Cr/Ti, Cr/V, W/Si, $WSi_2$/Si, and W/$B_4C$, which the minimum lateral gradual reaches 0.01 nm/mm with the total length of 70 mm.

As an example, the designed parameters of $WSi_2$/Si multilayer polarizer are given in **Table 3**. The working energy range is 430–760 eV. The calculated s- and p-reflectivities at every 5 mm are shown in **Figure 2**.

| Sample | | Position (mm) | | | | | | Energy range (eV) |
|---|---|---|---|---|---|---|---|---|
| | | L0 | L5 | L10 | L15 | L20 | L25 | |
| S1-[$WSi_2$/Si]$_{400}$ | Thickness (Å) | 15.33 | 16.23 | 17.23 | 18.25 | 19.31 | 20.30 | – |
| | Peak energy at 45° (eV) | 578.9 | 547.3 | 516.0 | 483.0 | 456.8 | 434.7 | 434.7–578.9 |
| S2-[$WSi_2$/Si]$_{400}$ | Thickness (Å) | 11.5 | 12.37 | 13.04 | 13.75 | 14.50 | 15.33 | – |
| | Peak energy at 45° (eV) | 764.2 | 715.0 | 679.0 | 644.6 | 611.7 | 578.9 | 578.9–764.2 |

**Table 3.** Designed parameters of $WSi_2$/Si lateral gradual multilayer.

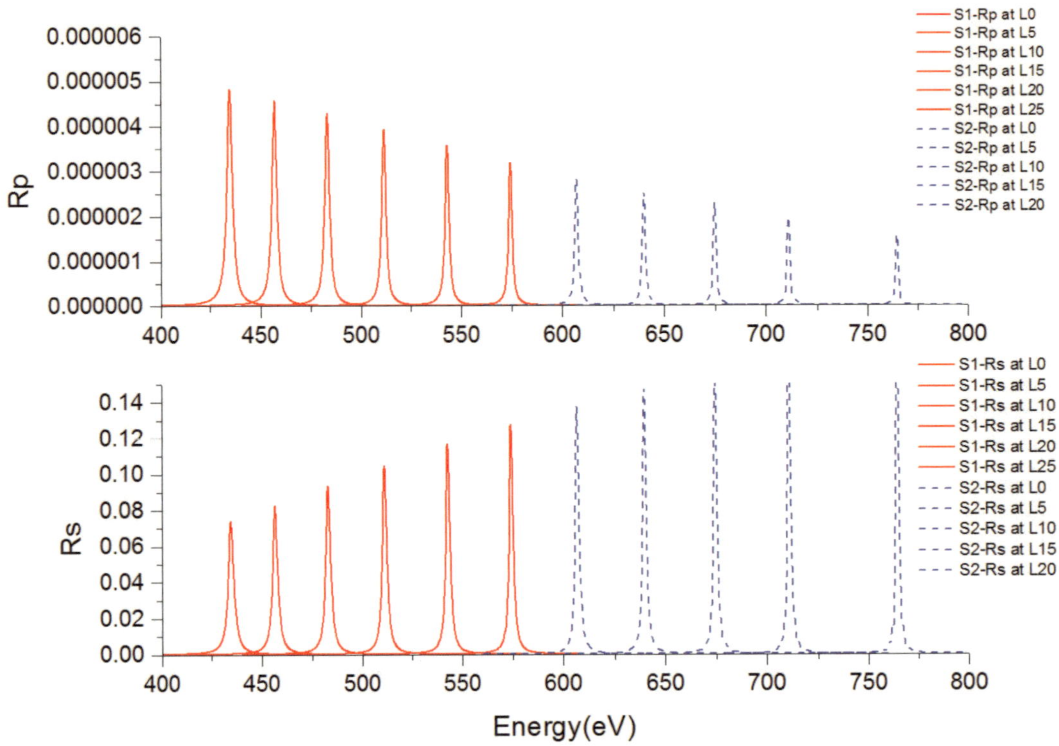


**Figure 2.** The calculated s- and p-reflectivities of $WSi_2$/Si lateral gradual multilayer.

## 3. Fabrication

The multilayer polarizers were deposited using a calibrated ultrahigh vacuum direct current (DC) magnetron sputtering system. They were deposited onto P(100) silicon wafer substrates at room temperature. Mask and speed control system were utilized in the fabrication of lateral gradual multilayers to control the lateral growth. The deposited multilayers were measured, for quality control, using a grazing incident angle X-ray reflectometry (XRR).

As an example, **Figure 3** shows the XRR curves for Mo/Si lateral gradual multilayer. Bragg peaks move evenly at different positions of the multilayer, which suggests a great homogeneity in lateral gradual variation. The appearances of all the fifth-order Bragg peaks prove the stability of fabrication technique.

The XRR result for $W/B_4C$ lateral gradual multilayer with 150 layers is shown in **Figure 4**. The measured period thicknesses at the positions of every 5 mm (every 2 mm at beginning) are given in **Table 4**. As a result, period thicknesses of $W/B_4C$ lateral gradual multilayer vary from 1.75 to 0.95 nm, and the multilayer covers the L absorption edges of Fe, Co, and Ni (600–900 eV) at quasi-Brewster angle, which is essential to the researches on magnetic materials.

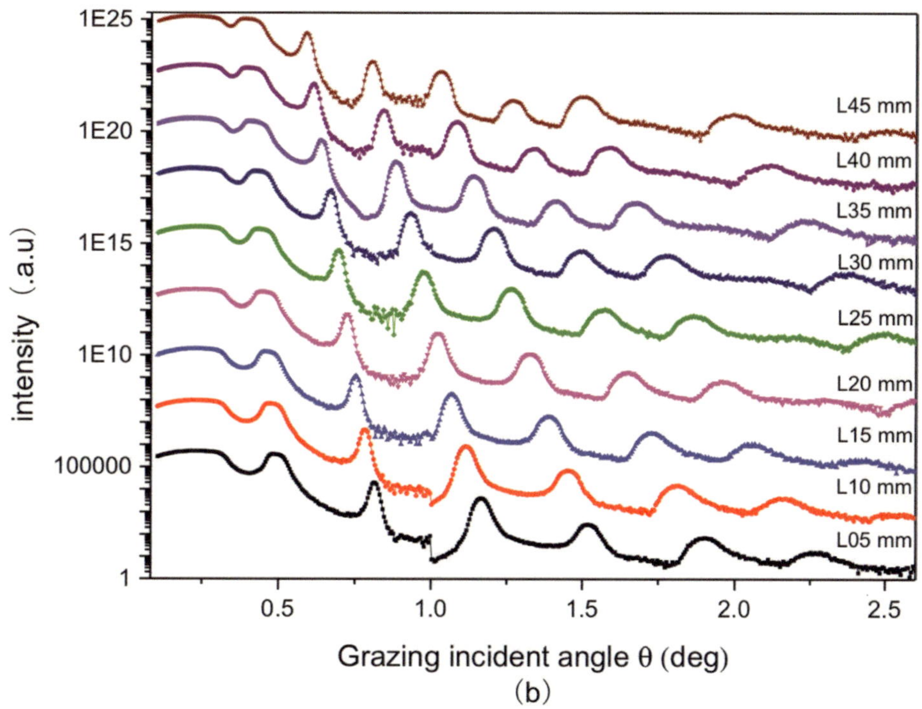


**Figure 3.** XRR results for Mo/Si lateral gradual multilayer at different positions (translations are set in *Y*-axis to make all the curves clear).

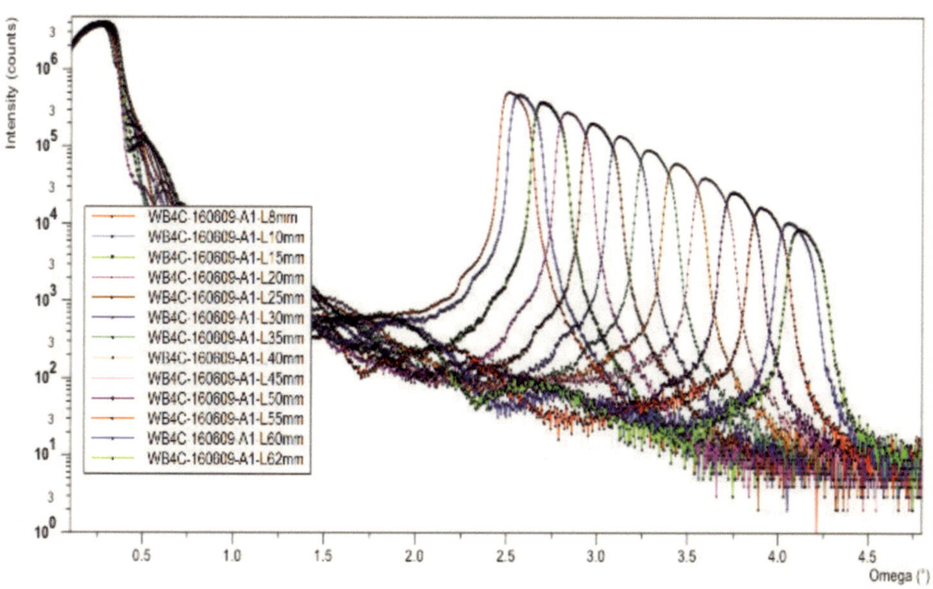


**Figure 4.** XRR results for $W/B_4C$ lateral gradual multilayer at different positions.

| Position (mm) | 08 | 10 | 15 | 20 | 25 | 30 | 35 | 40 | 45 | 50 | 55 | 60 | 65 |
|---|---|---|---|---|---|---|---|---|---|---|---|---|---|
| *D* (nm) | 1.75 | 1.73 | 1.63 | 1.56 | 1.48 | 1.40 | 1.34 | 1.28 | 1.22 | 1.17 | 1.13 | 1.08 | 0.95 |
| Energy at 45° (eV) | 501 | 507 | 538 | 562 | 592 | 626 | 654 | 685 | 719 | 749 | 776 | 812 | 923 |

**Table 4.** Measured periods of $W/B_4C$ lateral gradual multilayer at different positions.

## 4. Characterization and applications in synchrotron radiation measurement

The performances of Mo/Y and Mo/Si aperiodic multilayers have been evaluated using the high precision polarimeter on beamline UE56/1-PGM at BESSY II. The details can be found in our previous articles [7–11].

Taking as an example, measured mean s-reflectivities of Mo/Si transmission phase retarders with the bandwidth of 2.0 nm are 10.4% at 47° and the corresponding polarization efficiency is up to 98.9%, which is nearly four times of the periodic one. The analyzer also exhibits high s-reflectivity over the angular range of 10° at the fixed wavelength.

The polarization performance of Mo/Si transmission phase retarders is evaluated with broadband analyzer. A transmittance ratio $T_p/T_s$ of 5–23 from 12.5 to 13.2 nm is achieved and a constant phase shift of 50° at wavelength of 13.8–15.5 nm with the relative transmittance is 0.8–1.2.

Using this broadband Mo/Si polarizing system, the polarization of BESSY II UE56/1-PGM at 13.1 nm has been determined for the first time to our knowledge. As a result, the light of monochromatic radiation is nearly completely circularly, where the Stokes parameter is $|S3| \geq 0.94$, and the linear polarization is the largest with −0.96 for in-plane radiation [11].

A flexible soft X-ray multilayer polarimeter has been designed and built on beamline I06 at diamond Light Source. This allows polarization measurements using multilayers with unprecedented reproducibility and accuracy. The polarimeter with its pinhole assemblies and hexapod is easy to align to a beamline. More detailed description about this polarimeter can be seen in Ref. [18].

Experiments to characterize polarization properties of broadband lateral gradual multilayers have been performed. The designed energy range is from 434 to 764 eV at quasi-Brewster angle for $WSi_2/Si$ lateral gradual multilayer. The measurements are performed at every 2.5 mm along the horizontal direction. The corresponding multilayer periods are from 1.18 to 2.03 nm. The highest s-reflectivity reaches 2% in the wide energy range of 434–764 eV, i.e., a 330 eV-broadband polarizer has been achieved.

The optical properties of Mo/Si and Mo/Y lateral gradual multilayer are evaluated at NSRL in Hefei. The designed parameters are shown in **Tables 5** and **6**.

Reflectivities at different position of both Mo/Si and Mo/Y lateral gradual multilayer are measured at metrology station of NSRL in Hefei. As shown in **Figures 5** and **6**, the highest reflectivity of Mo/Si multilayer is more than 50% and obtained at the wavelength of 14.4 nm (86 eV). Mo/Y multilayer achieves the highest reflectivity of 32% at the wavelength of 13 nm (95 eV). The Bragg peaks of both multilayers distribute homogeneously in the energy-space at different positions, which shows a great stability in the gradient control during fabrication along the horizontal direction.

| Position (mm) | L05 | L10 | L15 | L20 | L25 | L30 | L35 | L40 | L45 |
|---|---|---|---|---|---|---|---|---|---|
| *D* (nm) | 11.89 | 12.45 | 13.07 | 13.72 | 14.37 | 15.15 | 16.05 | 17.03 | 18.06 |

**Table 5.** Designed parameters of Mo/Si lateral gradual multilayer.

| Position (mm) | L05 | L10 | L15 | L20 | L25 | L30 | L35 | L40 | L45 | L50 | L55 | L60 | L65 |
|---|---|---|---|---|---|---|---|---|---|---|---|---|---|
| *D* (nm) | 7.47 | 7.74 | 8.00 | 8.30 | 8.66 | 9.04 | 9.38 | 9.75 | 10.17 | 10.65 | 11.17 | 11.67 | 12.05 |

**Table 6.** Designed parameters of Mo/Y lateral gradual multilayer.

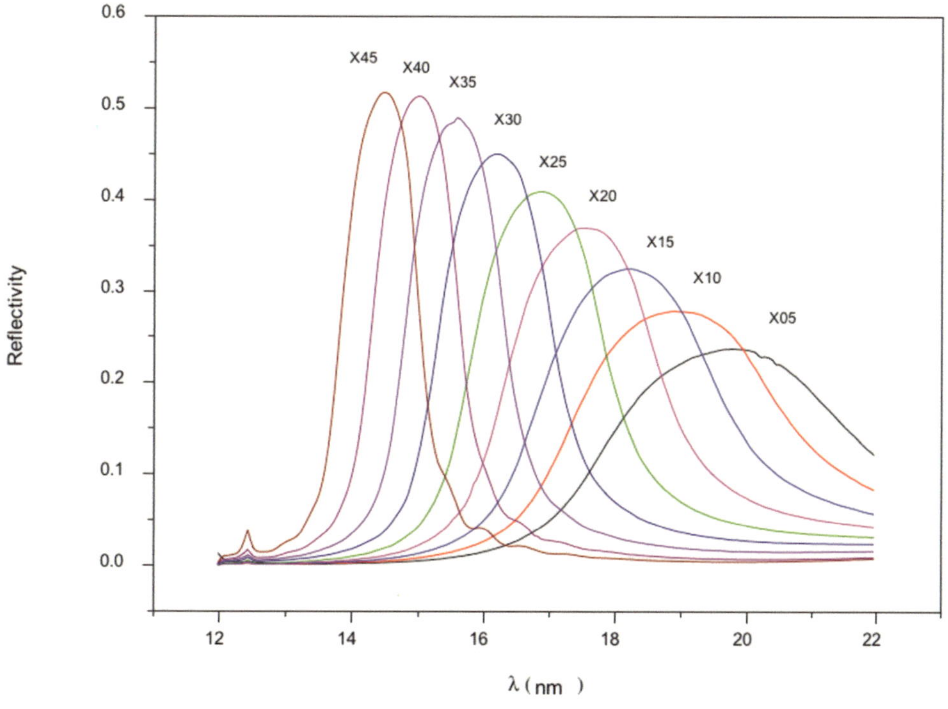


**Figure 5.** Reflectivities of Mo/Si lateral gradual multilayer at different positions.

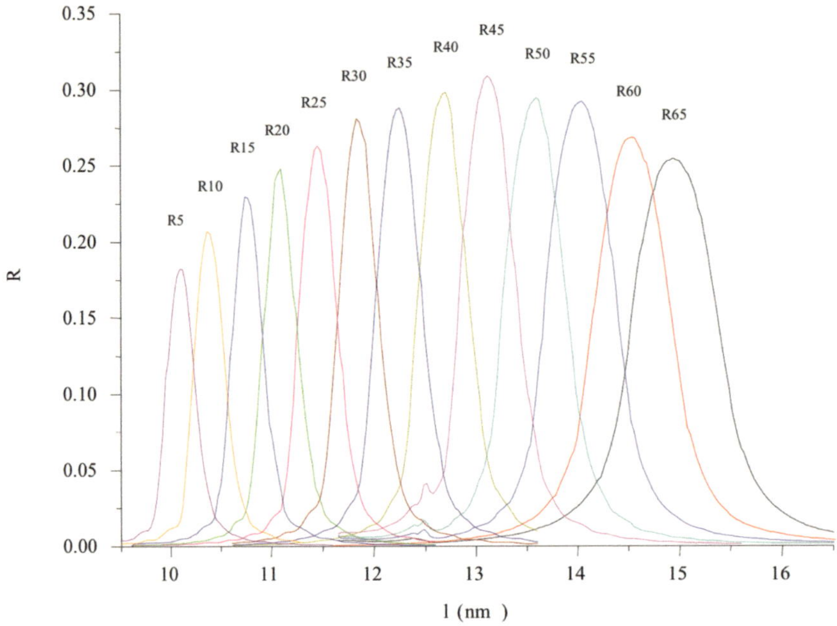


**Figure 6.** Reflectivities of Mo/Y lateral gradual multilayer at different positions.

## 5. Polarimeter in Beijing Synchrotron Radiation Facility (BSRF) and application

A compact high precision eight-axis automatism and two-axis manual soft-ray polarimeter has been designed, constructed, and installed in 3W1B at the Beijing Synchrotron Radiation Facility (BSRF). The beamline is a windowless full ultra-high vacuum (UHV) soft X-ray beamline utilizing a permanent-magnet wiggler source. It supplies focusing monochromatic soft X-rays. The photon energy region is from 50 to 1600 eV. There are four operational modes in this polarimeter device, which are double-reflection, double-transmission, front-reflection-behind-transmission, and front-transmission-behind-reflection. The schematic diagrams of four modes are shown in **Figure 7**. The chamber of the polarimeter links to the beamline by UHV ferrofluidics feedthrough on the upstream. It can be rotated around the axis of the incident beam. The chamber is supported on gimbals and installed in a rigid bench which can move in three dimensions. The schematic diagram and the photo of the polarimeter are shown in **Figures 8** and **9** [19].

The polarization properties of Mo/Y, Mo/$B_4C$, and W/$B_4C$ lateral gradual multilayer have been characterized by this polarimeter. Results are shown in **Figure 10**.

Mo/$B_4C$ lateral gradual multilayer polarizer is used to measure the polarization characteristics of the beamline 3W1B at BSRF. The measured parameters are given in **Table 7**. A highest s-reflectivity of 12% is achieved at the energy of 188 eV. **Figure 11** shows the measured reflectivities, where, p-reflectivities are not low enough comparing to s-reflectivities. That suggests the beam coming out from 3W1B is not completely linear polarized, which coincides with the result shown in **Table 7** that the average degree of polarization is around 72% for the beamline 3W1B.

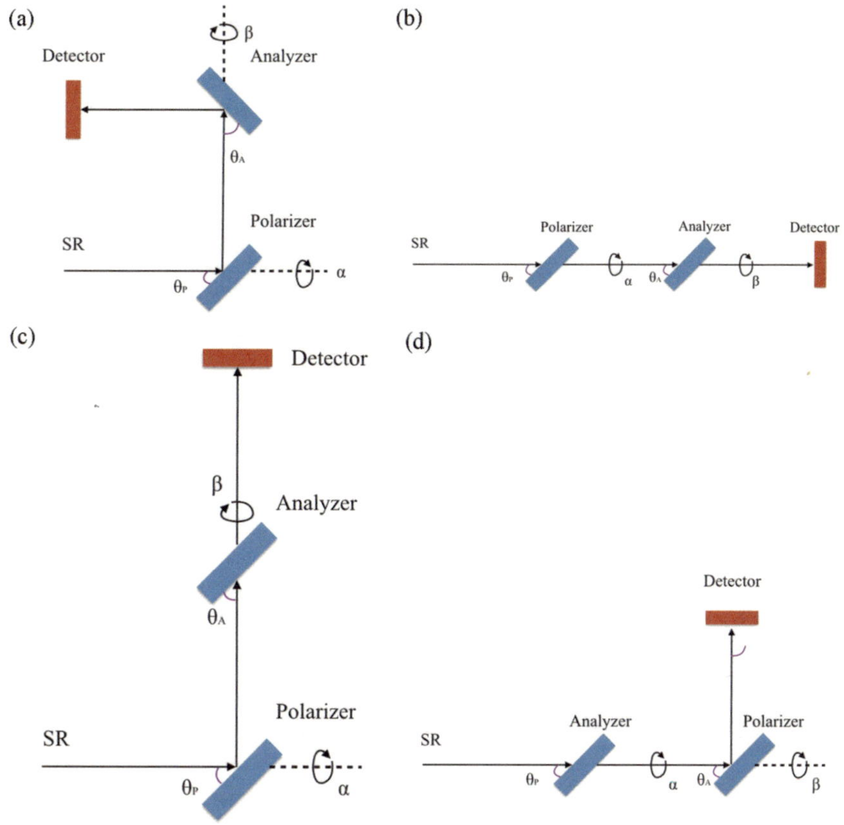


**Figure 7.** Schematic diagrams of four operational modes in polarimeter: (a) double-reflection; (b) double-transmission; (c) front-reflection-behind-transmission; (d) front-transmission-behind-reflection.

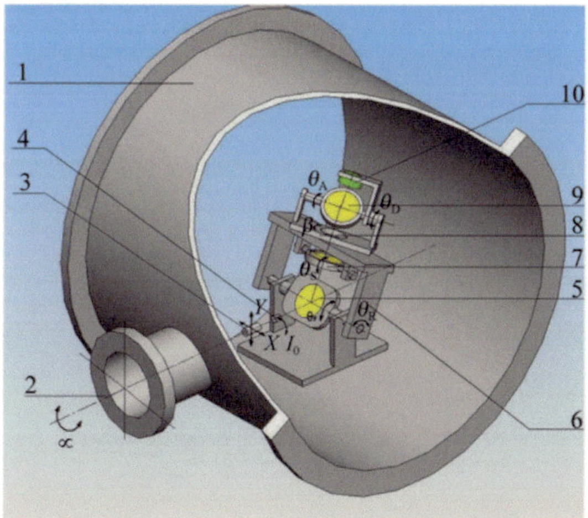


**Figure 8.** Schematic diagram of the polarimeter at BSRF: (1) chamber; (2) azimuthal angle of polarizer; (3) collimator translation x-y stage; (4) $I_0$ detector; (5) polarizer; (6) moving rocker; (7) sample stage; (8) rotational stage to change azimuth angle of analyzer; (9) analyzer; and (10) main detector.

**Figure 9.** Photo of the polarimeter device.

To obtain complete liner polarization at 3W1B, two twins-born Cr/C multilayers are used as polarizer and analyzer at the energy of 206 eV. Both Cr/C multilayers have the exactly same optical characteristic with bi-layer number of 100 and period thickness of 4.318 nm. Using the double-reflection mode of the soft X-ray polarimeter, a nearly complete linear polarized beam is achieved. The schematic diagram is the same as shown in **Figure 7(a)**. Both multilayers are set at their quasi-Brewster angles. As a result, the degree of linear polarization is raised from 0.585 to 0.995 after polarized by Cr/C multilayer at 206 eV. The detailed results of the polarization optimization at other energies are shown in **Table 8** [20].

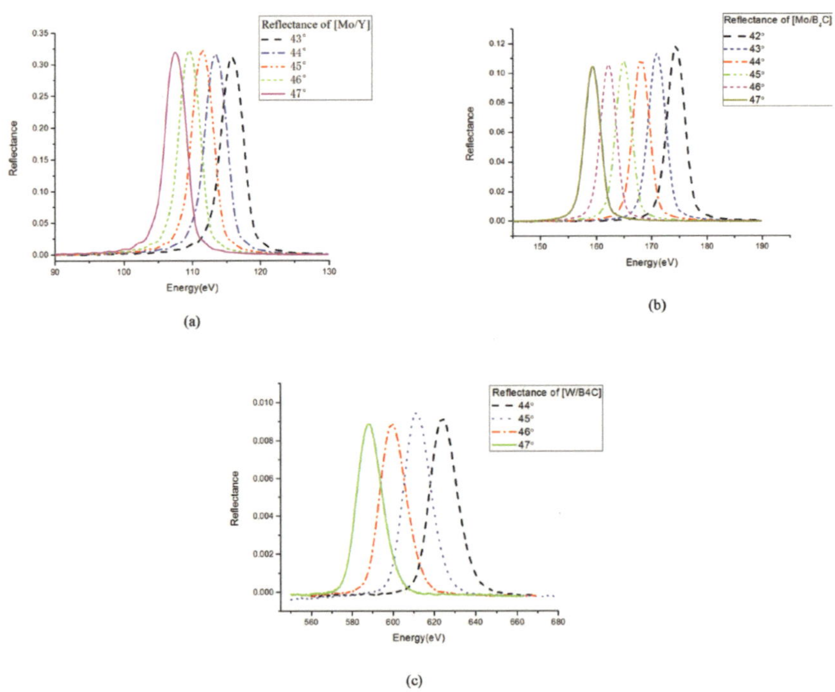


**Figure 10.** S-reflectivity as a function of energy: (a) Mo/Y; (b) $Mo/B_4C$; (c) $W/B_4C$.

| Position (mm) | $I_s/I_p$ | $I_s-I_p/I_s+I_p$ | Peak energy (eV) $R_s/R_p$ | $R_s$ | $R_s/R_p$ |
|---|---|---|---|---|---|
| L9 | 5.98 | 0.71 | 187.7/187.6 | 0.12 | 5.98 |
| L26 | 6.55 | 0.74 | 163.4/163.2 | 0.10 | 6.54 |
| L44 | 6.80 | 0.74 | 140.7/140.5 | 0.10 | 6.80 |
| L61 | 6.68 | 0.74 | 121.7/122.4 | 0.09 | 6.69 |

**Table 7.** Measured parameters of $Mo/B_4C$ lateral gradual multilayer at different positions.

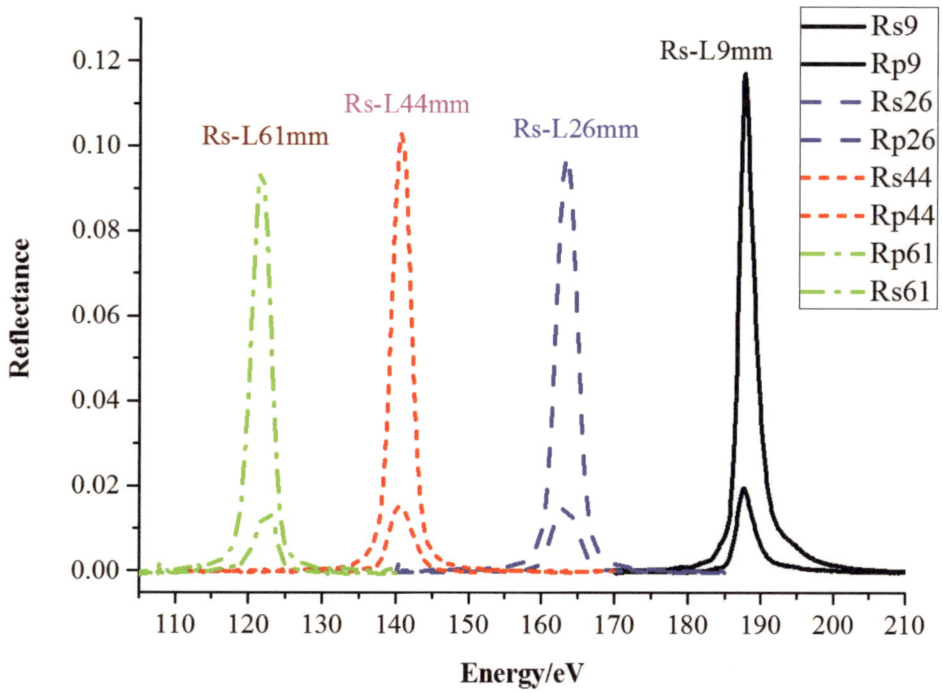


**Figure 11.** S-reflectivities and p-reflectivities at different energies of the beamline 3W1B.

| Energy (eV) | $P_L$ (linear polarization degree) before polarizing | $P_L$ (linear polarization degree) after polarizing |
|---|---|---|
| 206.0 | 0.585 | 0.995 |
| 92.5 | 0.443 | 0.99 |
| 77.5 | 0.373 | 0.985 |
| 65.0 | 0.400 | 0.950 |

**Table 8.** Measurement results of linear polarization at different energies.

## 6. Conclusion

The multilayers can be utilized in EUV and soft X-ray polarization applications. A series of multilayers with periodic and aperiodic thicknesses are designed to broaden the bandwidth. Measurements are performed at BESSY-II, Diamond and Beijing Synchrotron Radiation Facility. For Mo/Y aperiodic multilayer, polarization efficiency is up to 20% over a 90–120 eV (10.5–13.5 nm) energy range. Mo/Si broadband transmission phase retarder is also realized. Combination using the polarimeter, complete polarization characteristics of the synchrotron radiation beam is performed. The results show that broadband aperiodic multilayer greatly simplifies polarization experiments. For further development of multilayer polarizers in soft X-ray region, especially the "water window" range, a series of lateral gradual multilayers are utilized. The performance of $WSi_2$/Si lateral gradual multilayer polarizer is evaluated at diamond light source, which, as a result, covers the energy range of 434–764 eV. In addition, polarization measurements of Mo/Si, Mo/Y, $Mo/B_4C$, $W/B_4C$, and other lateral gradual multilayers have been performed at Beijing Synchrotron Radiation Facility and National Synchrotron Radiation Laboratory in Hefei. These researches create some novel experimental methods, such as soft X-ray magnetic circular dichroism, element resolution Faraday effect, and Kerr effect measurements, which provide the powerful tools for biology, medicine, information, material science, etc.

## Acknowledgements

The authors are indebted to Dr. Igor V. Kozhevnikov for useful discussions. Special thanks are due to Dr. Franz. Schäfers, Dr. Hongchang Wang, Matthew Hand, Hongjun Zhou, Tonglin Huo, and Dr. Lijuan Sun for their kind assistance during the measurement at BESSY, Diamond, NSRL, and BSRF. The authors are grateful to Dr. Alan Michette, Dr.Keith Powell, Dr. Slawka Pfauntsch, Dr. Sawhney Kawal, and Prof. Zhanshan Wang for their great support during this work. The author would also like to thank Miao Li, Shengming Zhu, Bin Ji, Jinwen Chen, and Jiayi Zhang for the participations during multilayer design, fabrication, and measurement processes. This work was supported by the National Natural Science Foundation of China (contract numbers 11375131 and U1432244), by the Royal Society—NSFC International Exchanges funding (reference IE141043 and 11511130051), and by the Fundamental Research Funds for Central Universities.

## Author details

Jingtao Zhu[1]* and Mingqi Cui[2]

*Address all correspondence to: jtzhu@tongji.edu.cn

1 MOE Key Laboratory of Advanced Micro-Structured Materials, School of Physics Science and Engineering, Tongji University, Shanghai, China

2 Beijing Synchrotron Radiation Facility, Institute of High Energy Physics, Chinese Academy of Sciences, Beijing, China

## References

[1] Wu Y, Parkin SSP, Stöhr J, et al. Direct observation of oscillatory interlayer exchange coupling in sputtered wedges using circularly polarized x rays. Applied Physics Letters. 1993;**63**(2):263-265

[2] Hoobler RJ, Korlahalli R, Boltich E, Serafin J. Characterization of interfacial layers and surface roughness using spectroscopic reflectance, spectroscopic ellipsometry, and atomic force microscopy. Proceedings of SPIE—The International Society for Optical Engineering. 2002;**4689**:756-764

[3] Hirai Y, Takahashi H, Komuro M, et al. Soft X-ray polarization and magnetic circular dichroism measurement system. Journal of Electron Spectroscopy & Related Phenomena. 1996;**80**:385-388

[4] Bahrdt J, Gaupp A, Gudat W, et al. Circularly polarized synchrotron radiation from the crossed undulator at BESSY. Review of Scientific Instruments. 1992;**63**(1):339-342

[5] Kortright JB, Rice M, Kim SK, et al. Optics for element-resolved soft X-ray magneto-optical studies. Journal of Magnetism & Magnetic Materials. 1998;**191**(1-2):79-89

[6] Kjornrattanawanich B, Bajt S, Seely JF. Mo/$B_4C$/Si multilayer-coated photodiode with polarization sensitivity at an extreme-ultraviolet wavelength of 13.5 nm. Applied Optics. 2004;**43**(5):1082

[7] Wang Z, Wang H, Zhu J, et al. Extreme ultraviolet broadband Mo/Y multilayer analyzers. Applied Physics Letters. 2006;**89**(24):241120-241120-3

[8] Wang Z, Wang H, Zhu J, et al. Broadband multilayer polarizers for the extreme ultraviolet. Journal of Applied Physics. 2006;**99**(5):056108-056108-3

[9] Wang ZS, Wang HC, Zhu JT, et al. Broad angular multilayer analyzer for soft X-rays. Optics Express. 2006;**14**(6):2533-2538

[10] Wang Z, Wang H, Zhu J, et al. Broadband Mo/Si multilayer transmission phase retarders for the extreme ultraviolet. Applied Physics Letters. 2007;**90**(3):054437

[11] Wang Z, Wang H, Zhu J, et al. Complete polarization analysis of extreme ultraviolet radiation with a broadband phase retarder and analyzer. Applied Physics Letters. 2007;**90**(8):79

[12] Yanagihara M, Maehara T, Nomura H, et al. Performance of a wideband multilayer polarizer for soft x rays. Review of Scientific Instruments. 1992;**63**(1):1516-1518

[13] Kortright JB, Rice M, Franck KD. Tunable multilayer EUV/soft x-ray polarimeter. Review of Scientific Instruments. 1995;**66**(2):1567-1569

[14] Kozhevnikov IV, Bukreeva IN, Ziegler E. Design of X-ray supermirrors. Nuclear Instruments & Methods in Physics Research. 2001;**460**(2):424-443

[15] Vidal B, Vincent P. Metallic multilayers for x rays using classical thin-film theory. Applied Optics. 1984;**23**(11):1794

[16] Kim DE, Lee SM, Jeon I. Transmission characteristics of multilayer structure in the soft x-ray spectral region and its application to the design of quarter-wave plates at 13 and 4.4 nm. Journal of Vacuum Science & Technology A Vacuum Surfaces & Films. 1999;**17**(2):398-402

[17] Kortright JB. Polarization properties of multilayers in the EUV and soft x ray. Proceedings of SPIE—The International Society for Optical Engineering. 1994;**2010**:160-167

[18] Wang H, Dhesi SS, Maccherozzi F, et al. High-precision soft x-ray polarimeter at Diamond Light Source. Review of Scientific Instruments. 2011;**82**(12):123301

[19] Sun L, Cui M, Zhu J, et al. Multilayer-based soft X-ray polarimeter at the Beijing Synchrotron Radiation Facility. Chinese Physics C. 2013;**37**(7):83-87

[20] Cui M, Sun L, Zhu J, et al. The soft X-ray polarimeter and applications at BSRF. Nuclear Techniques. 2010;**1234**(1):641-644

Chapter 5

# Spectroscopic Ellipsometry Study of Organic-Inorganic Halide: $FAPbI_xBr_{3-x}$ Perovskite Thin Films by Two-Step Method

Hajime Shirai

Additional information is available at the end of the chapter

http://dx.doi.org/10.5772/intechopen.70281

**Abstract**

Spectroscopic ellipsometry (SE) was used to investigate the role of isopropyl alcohol (IPA) solvent in the synthesis of organic-lead-halide perovskite $CH(NH_2)_2PbI_xBr_{3-x}$ [$FAPbI_xBr_{3-x}$] thin films including the effect of I/Br composition ratio by the two-step reaction of an amorphous (a-)$PbI_xBr_{2-x}$ layer and $FAI_xBr_{1-x}$ solution diluted in IPA. An optical dispersion model was developed to extract the complex refractive index $N$ (=$n$ + $ik$), optical transition, and film thickness of $FAPbI_xBr_{3-x}$ perovskites by SE analysis at different I/Br composition ratio as a function of immersion time in a solution of $FAI_xBr_{1-x}$ diluted in IPA. SE combined with X-ray diffraction (XRD), Fourier-transform infrared spectroscopy (FTIR), and scanning electron microscopy (SEM) revealed that Br incorporation into films promoted the densification of $FAPbI_xBr_{3-x}$ perovskite network along with increased film thickness and volume fraction of void. IPA promoted film crystallization of a-$PbI_xBr_{2-x}$ accompanied by the formation of surface roughness, grain boundaries, and voids, followed by enhanced diffusion of $FAI_xBr_{1-x}$ into the grain boundaries/voids in the mesoporous crystallized $PbI_xBr_{2-x}$ network. These processes contribute synergistically to the growth of the perovskite structure.

**Keywords:** ellipsometry, organic-lead-halide perovskite, two-step reaction, role of isopropyl alcohol (IPA) solvent, growth mechanism

## 1. Introduction

In recent years, organic-inorganic halide perovskite solar cells have significantly attracted much interest as high-efficiency solar cell material due to their high absorption coefficient, ambipolar carrier transport, and long-carrier diffusion lengths, which have resulted in power conversion efficiencies (PCEs) as high as 20.1% [1–3]. Methylammonium [$(CH_3NH_3)^+$; MA] lead triiodide ($MAPbI_3$), which is used in the most efficient perovskite solar cells, has a bandgap ($E_g$) of around 1.57 eV [4, 5], a very sharp absorption edge [6], and an impressively low difference between the open-circuit voltage ($V_{oc}$) and its bandgap potential ($E_g/q$) [6–8]. However, the instability of $MAPbI_3$ with respect to humidity, temperature, and light soaking is a serious problem hindering long-term reliability of the photovoltaic performance. Therefore, the formamidinium [$(NH_2)_2CH^+$; FA] lead triiodine ($FAPbI_3$) perovskite, where the ionic radius of the A-site $FA^+$ ion (253 pm) is larger than that of $MA^+$ (217 pm), has recently been extensively studied due to better-phase stabilization and higher photovoltaic performance over other systems such as $MAPbI_3$, and better stability for air storage with an extension of the optical absorption edge to the infrared region up to around 840 nm [5]. More recently, interest has extended to the $FAPbI_xBr_{3-x}$ and $Cs_{0.2}FA_{0.8}PbI_xBr_{3-x}$ perovskite, which has even higher stability against humidity as well as temperature. A bandgap is also tunable from 1.43 to 2.35 eV by substitutional doping of bromide ions to iodine sites [9].

In addition, various methods for the fabrication of perovskite thin films have been extensively studied, such as a one-step solution process [10], sequential two-step reaction [11, 12], and vacuum deposition [13]. Although an efficient PCE of around 20% has been achieved using uniform and dense perovskite layers prepared by a one-step solution process [14, 15], this involves complex procedures such as precise control of the composition ratio in the precursor solution and the rate of evaporation to adjust the composition ratio of the perovskite film. An alternative approach is a two-step deposition method using porous $PbX_2$ (X = I, Cl, Br) films and organic halide solutions such as MAX and FAX diluted in 2-isopropanol alcohol [IPA; $(CH_3)_2COH$] solvent on an electron transport layer such as mesoporous $TiO_2$ or ZnO, which results in better coverage and uniformity than the one-step process. However, the film morphology of the porous $PbX_2$ in the two-step deposition method has a strong effect on the chemical reactivity with the alkyl amine halide, which in turn determines the film homogeneity and crystallinity of the resultant perovskite thin films. To control the morphology of $PbX_2$ films, solvent vapor treatment [16], and additives such as hydriodic acid [17], dimethyl sulfoxide (DMSO), and *N*-methyl-2-pyrrolidone [18] have been tested. For example, Miyadera et al. reported that the rate of perovskite formation from polycrystalline $PbI_2$ on mesoporous $TiO_2$ was approximately one order of magnitude faster than that of well-oriented dense $PbI_2$ on glass [19]. In addition, the type of solvent is sensitive to the morphology of $PbI_2$. *N,N*-Dimethylformamide (DMF; $CH_3NCHO$) is generally used to promote the crystallization of $PbI_2$. The use of DMSO or cumene hydroperoxide (CHP; $C_6H_5C(CH_3)_2OOH$), which has lower vapor pressures than DMF, has been reported to suppress the rapid crystallization and to form an amorphous structure that reacts preferentially with MAI diluted in IPA solvent, resulting in a slower rate of crystallization of $MAPbI_3$ and a more homogeneous film structure [20, 21]. However, the role of the polar IPA solvent in the synthesis of $FAPbI_xBr_{3-x}$ perovskites is still not clear because the reaction rate is very fast. Thus, to understand the effect of Br incorporation on the fine structure and the growth mechanism of perovskites by a two-step method from $PbI_xBr_{2-x}$ film and $FAI_xBr_{1-x}$ solution diluted in IPA solvent, a slower reaction rate is preferable from amorphous (a-)$PbI_xBr_{2-x}$ rather than from mesoporous crystalline $PbX_2$ by immersion in FAX/IPA solution at lower temperatures.

In this chapter, we present the role of the IPA solvent in the synthesis of $FAPbI_xBr_{3-x}$ perovskites including the effect of Br incorporation in the films by a two-step solution process using spectroscopic ellipsometry (SE) as well as X-ray diffraction (XRD), Fourier-transform infrared spectroscopy (FTIR), and scanning electron microscopy (SEM) observation.

## 2. Experimental section

A-$PbI_xBr_{2-x}$ thin films with a film thickness of around 150 nm were prepared on glass by spin coating a mixture of $FAI_xBr_{1-x}$ and $PbI_xBr_{2-x}$ powder ($PbI_2$:$PbBr_2$ = 3:1 and 3:2 molar ratio) in a DMF/CHP (95:5, v/v) cosolvent at 3000 rpm for 30 s. A-$PbI_xBr_{2-x}$ films prepared using the DMF/CHP cosolvent promoted the growth of perovskite structure after immersion in $FAI_xBr_{1-x}$/IPA solution, whereas no reaction to form the perovskite occurred in a-$PbI_xBr_{2-x}$ films prepared from 100 vol% DFM solvent. Thus, a-$PbI_xBr_{2-x}$ fabricated with DMF/CHP cosolvent was used as a precursor to promote the formation of perovskite structure. A-$PbI_xBr_{2-x}$ films were immersed in a solution of $FAI_xBr_{1-x}$ (FAI:FABr = 3:1 molar ratio) in IPA solvent for various times ($t_{im}$) at 25°C. Finally, the $FAPbI_xBr_{3-x}$ films were rinsed with IPA to remove the residual solution and contaminants from the film surface.

The refractive index $n$ and extinction coefficient $k$ of the corresponding $FAPbI_xBr_{3-x}$ films were determined using a phase-modulated spectroscopic ellipsometer (UVISEL, Horiba Jobin Yvon). The ellipsometric angles, Ψ and Δ, which determine the complex reflection coefficient ratio, $\rho = \tan\Psi e^{i\Delta}$, were also measured for 71 points in the range of 1.5–5.0 eV (0.05 eV step) with an integration time of 200 ms at each photon energy. The angle of incidence, $\theta_i$, of the probe light was 70°. The measured values for the $n$ and $k$ spectra, denoted by <*n*> and <*k*>, include information regarding the surface roughness, voids, and bulk inhomogeneity of the perovskite film and the $PbI_xBr_{2-x}$ precursor. The bulk component of optical constants for the perovskite films, $n$ and $k$, and the volume fractions of perovskite and voids ($f_{perov}$, $f_{void}$) were determined using a modified amorphous model based on the Forouhi-Bloomer (FB) model described in Eq. (1), combined with the Bruggeman effective medium approximation (EMA) [22, 23]:

$$n(\omega) = \frac{B \cdot (\omega - \omega_j) + C}{(\omega - \omega_j)^2 + \Gamma_j^2} \tag{1}$$

$$k(\omega) = \begin{cases} \dfrac{A_j \cdot E_0 \cdot (\omega - \omega_g)^2}{(\omega - \omega_g)^2 + \Gamma^2} & for\ \omega > \omega_g \\ 0 & for\ \omega > \omega_g \end{cases} \tag{2}$$

where

$$\begin{aligned} B &= \frac{A_j}{\Gamma_J} \cdot [\Gamma_j^2 - (\omega_j - \omega_g)^2] \\ C &= 2 \cdot A_j \cdot \Gamma_j \cdot (\omega_j - \omega_g) \end{aligned} \tag{3}$$

The quantities $A_j$, $\Gamma_0$, $\omega_j$, and $\omega_g$ represent the resonance amplitude, the oscillator-broadening parameter, the resonance energy, and the Tauc optical gap, respectively. Spectral fitting was performed using the optical constant for bulk $FAPbI_xBr_{3-x}$ perovskite described in Eq. (1), and $f_{perov}$, $f_{void}$, and each film thickness $d_i$ ($i$ = 1, 2, 3), as variables. A probable structure was determined using the optical model by minimization of the standard deviation $\chi^2$ between the measured and calculated ellipsometric error parameters using a linear regression method as follows [24]:

$$\chi^2 = \frac{1}{2N-M}\sum_{i=1}^{N}\left[\left(\tan\Psi_i^c - \tan\Psi_i^m\right)^2 - \left(\cos\Delta_i^c - \cos\Delta_i^m\right)^2\right] \quad (4)$$

where the superscripts $c$ and $m$ represent the calculated and measured values, and $N$ and $M$ are the numbers of the measured and calculated wavelengths, respectively.

The fine structures of the a-$PbI_xBr_{2-x}$ and perovskite $FAPbI_xBr_{3-x}$ thin films were also evaluated using X-ray diffraction (Bruker D8 ADVANCE ECO) with the diffractometer in Bragg-Brentano mode using Cu K$\alpha$ radiation with a Lynxeye XE detector, Fourier-transform infrared spectroscopy (Shimadzu spectrometer, IRTracer-100), field emission scanning electron microscopy (FESEM; S4800, Hitachi High Technologies, Japan) with energy-dispersive X-ray spectroscopy (EDX) (Bruker XFlash 5030/Quantax 400), and atomic force microscopy (AFM; NanonavieII/SPI-3800, Hitachi High-Tech Science, Japan) to support the results of the SE analysis.

## 3. Results and discussion

### 3.1. Characterization of $FAPbI_xBr_{3-x}$ perovskite films

**Figure 1a** shows XRD patterns of corresponding $FAPbI_xBr_{3-x}$ films on glass synthesized by the two-step method using a-$PbI_xBr_{2-x}$ layer and $FAI_xBr_{1-x}$ solution in IPA at various $t_{im}$. No XRD diffraction peaks that would indicate a crystalline phase were observed for spin-coated

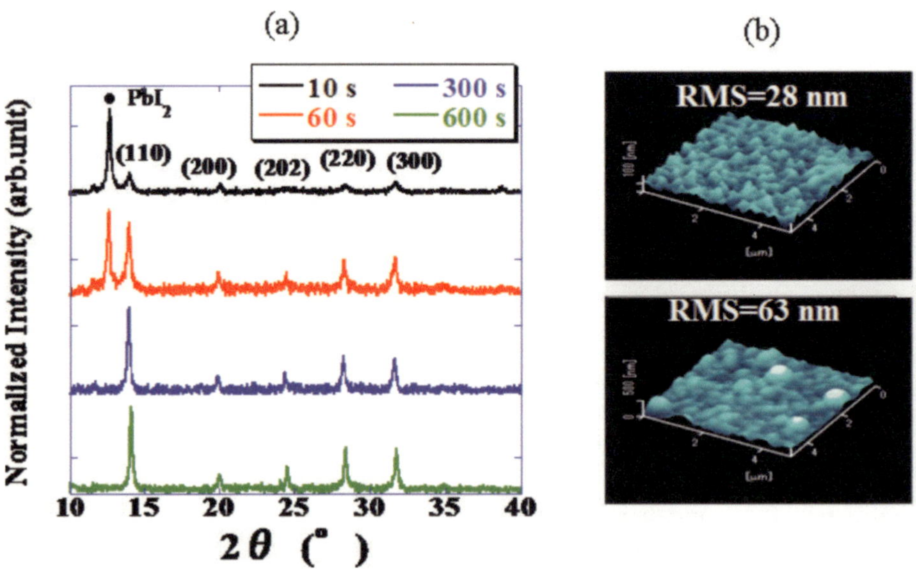


**Figure 1.** (a) XRD pattern of $FAPbI_xBr_{3-x}$ films synthesized at various $t_{im}$. (b) AFM image of $t_{im}$ of 10- and 600-s-immersed films.

http://dx.doi.org/10.5772/intechopen.70281

$PbI_xBr_{2-x}$, which suggests that the film structure was mainly amorphous. However, the film immersed for 10 s showed a diffraction peak at $2\theta$ = 12.6°, which is attributed to the (1 0 0) crystal orientation in $PbI_xBr_{2-x}$, and peaks were observed at $2\theta$ =14.6° and 28°, which correspond to the (1 1 0) and (2 2 0) diffraction peaks of the $FAPbI_xBr_{3-x}$ perovskites, respectively. A trace of the hexagonal structure δ phase was also indicated by the peak at $2\theta$ = 11.7°, which is undesirable with regard to the photovoltaic performance. The diffraction peaks attributed to the $FAPbI_xBr_{3-x}$ perovskites increased with suppression of the δ phase peaks and the diffraction peak at $2\theta$ =12.6° attributed to the (1 0 0) crystal orientation in $PbI_xBr_{2-x}$ when $t_{im}$ was increased up to 600 s. The AFM image of 10 and 600 s immersed film is also shown in **Figure 1b**. RMS is increased from 28 to 63 nm. These results suggest that the I/Br composition ratio determines the fine structure, and the formation of a perovskite structure is promoted together with an increase of the surface roughness, grain boundaries, and voids.

### 3.2. SE study of perovskite $FAPbI_xBr_{3-x}$ films

**Figure 2** shows <*n*> and <*k*> spectra for the $FAPbI_xBr_{3-x}$ perovskite films prepared at various $t_{im}$, including those for the a-$PbI_xBr_{2-x}$ precursor film. Once the a-$PbI_xBr_{2-x}$ layers were immersed in the $FAI_xBr_{1-x}$ solution in IPA, the <*n*> value decreased gradually, which suggests an increase in porosity, voids, and/or surface roughness in the a-$PbI_xBr_{2-x}$ structure. By contrast, the <*k*> value that corresponds to optical absorption $\alpha = 4\pi k/\lambda$ ($\alpha = 4\pi k/\lambda$) in the visible to near-infrared region was markedly increased for the film immersed for 10 s, followed by a decrease with $t_{im}$ up to 600 s. Thus, the time evolutions of <*n*> and <*k*> spectra with $t_{im}$ include information of the growth chemistry of the $FAPbI_xBr_{3-x}$ perovskite structure by the two-step process of a-$PbI_xBr_{2-x}$ layer formation and immersion in the $FAI_xBr_{1-x}$ solution in IPA.

**Figure 3a** shows the <*n*> and <*k*> spectra for the spin-coated a-$PbI_xBr_{2-x}$ layer included in **Figure 2** together with the best-fit calculated spectra. A single-layer model consisting of a modified FB model was used to determine the optical constant of the bulk component of the a-$PbI_xBr_{2-x}$ layer and the film thickness. The best-fit <*n*> and <*k*> spectra are also included as dotted lines. Good-fitting results were obtained over the entire energy region from 1.5 to 5 eV

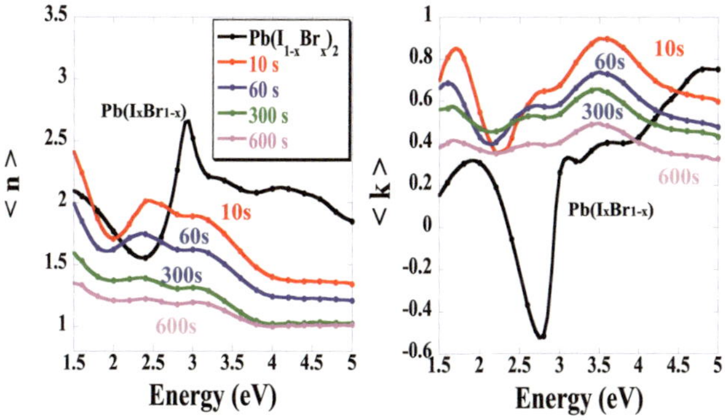


**Figure 2.** Measured <*n*> and <*k*> spectra for $FAPbI_xBr_{3-x}$ perovskite films prepared on glass by the immersion of a-$PbI_xBr_{2-x}$ in a solution of $FAI_xBr_{1-x}$ in IPA solvent at various $t_{im}$.

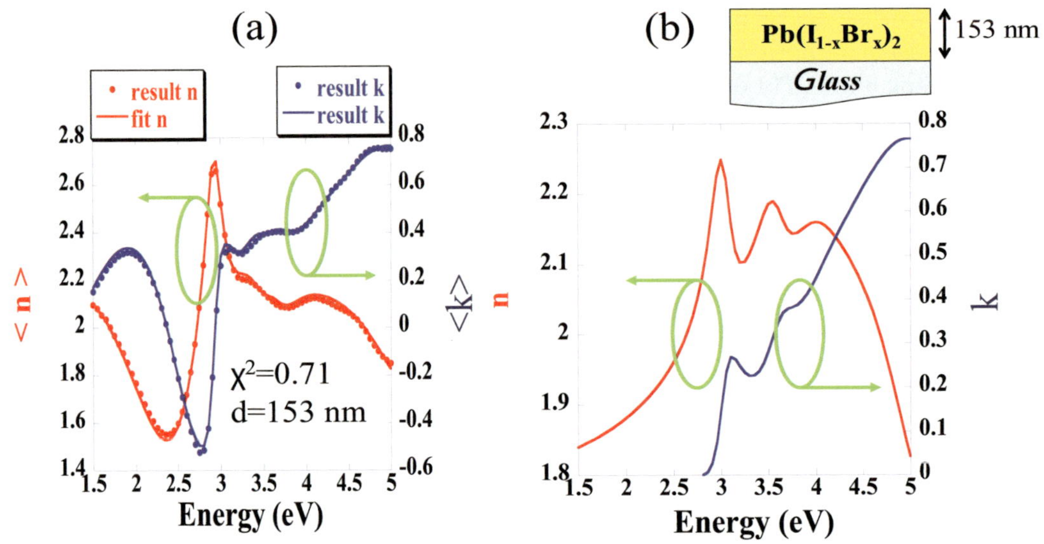


**Figure 3.** (a) Measured <*n*> and <*k*> spectra for spin-coated a-$PbI_xBr_{2-x}$ included in **Figure 2** together with the best-fit calculated spectra. (b) The *n* and *k* spectra for bulk component of a-$PbI_xBr_{2-x}$ extracted by a spectral-fitting procedure using a single-layer model.

with $\chi^2$ of 0.71. The *n* and *k* spectra of the a-$PbI_xBr_{2-x}$ bulk component along with the optical model used for the spectra fitting are shown in **Figure 3b**. The film thickness was 153 nm, which was almost consistent with that determined by FESEM observation and a thickness profiler. The fine structures were observed at 3.1, 3.7, and 4.2 eV in a-$PbI_xBr_{2-x}$, which are analogous to those at 2.95, 3.50, and 3.9 eV attributed to the respective $\Gamma_4^+ \rightarrow \Gamma_4^-(p_{xy})$, $\Gamma_4^+ \rightarrow \Gamma_4^-(p_z)$, and $A_4^+ \rightarrow A_4^-$ (conduction band) optical band transitions in $PbI_2$ [25]. These high-energy shifts in the fine structures are due to the addition of Br.

To understand the formation mechanism of perovskite structure using a two-step method, the measured <*n*> and <*k*> spectra were analyzed using two- or three-layer models consisting of a perovskite $FAPbI_xBr_{3-x}$ top layer and a $PbI_xBr_{2-x}$ underneath layer using constant (*n*, *k*) spectra for $FAPbI_xBr_{3-x}$ and $PbI_xBr_{2-x}$, and $f_{perov}$, $f_{Pb(I,Br)2}$, $f_{void}$, and $d_i$ as variables. However, the (*n*, *k*) spectra calculated using several optical models deviated far from the measured spectra, which suggests that modelization of the optical constant for the transitional stage from a-$PbI_xBr_{2-x}$ to the $FAPbI_xBr_{3-x}$ perovskite structure should be considered at each $t_{im}$. Among single-layer to four-layer optical models, the three-layer model showed the best fit with four Lorentzian oscillators for a $PbI_xBr_{2-x}$ and $FAPbI_xBr_{3-x}$ composite phase including $f_{perov}$, $f_{void}$, and $d_i$ ($i$ = 1, 2, 3) as variables.

**Figure 4a** shows the best-fit *n* and *k* spectra for the $FAPbI_xBr_{3-x}$ perovskite bulk component prepared at various $t_{im}$. The *n* value increased in the first 10 s and then decreased with increasing $t_{im}$, whereas the *k* values increased with the appearance of several fine structures attributed to the optical band transition of $FAPbI_xBr_{3-x}$. The absorption edge shifted to a lower energy from 3.1 to 1.63 eV for 60-s immersion and then became almost independent of $t_{im}$. In

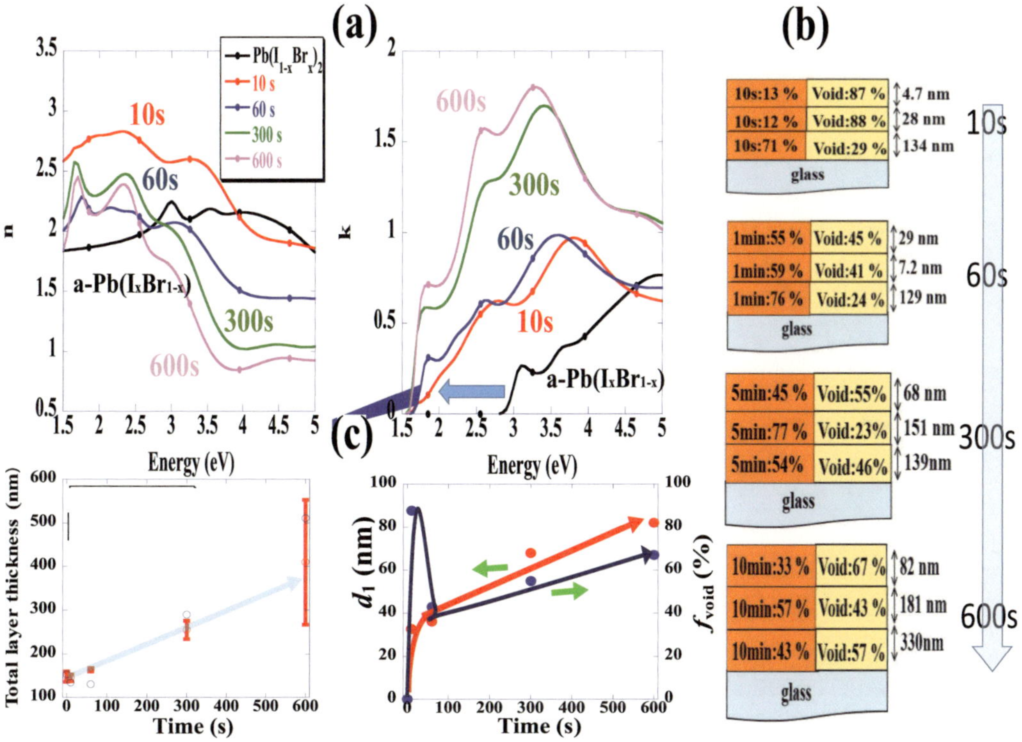


**Figure 4.** (a) Best-fit $n$ and $k$ spectra for the $FAPbI_xBr_{3-x}$ perovskite bulk component for corresponding films prepared at various $t_{im}$. (b) Best-fit optical models for corresponding films at various $t_{im}$. (c) Total film thickness $d$, $d_1$, and $f_{void}$ in the top layer plotted as a function of $t_{im}$.

addition, fine structures at 2.55 and 3.8 eV were shifted to the infrared region with an increase in their amplitude when $t_{im}$ was increased up to 600 s. In particular, the layer immersed for 600 s exhibited four fine structures in the $k$ spectra at 1.63, 2.52, 3.25, and 4.79 eV, which were attributed to the optical transition of the perovskite structure.

The best-fit optical models for the various $t_{im}$ are summarized in **Figure 4b**, including the corresponding $\chi^2$ values. For the a-$PbI_xBr_{2-x}$ films immersed for 10 and 60 s, no significant differences in $f_{perov}$ and $f_{void}$ were observed between the two- and three-layer models. The $f_{perov}$ and $f_{void}$ values were almost identical for both the top and intermediate layers, despite the use of the three-layer model. A prominent feature was that $f_{void}$ was increased significantly for the entire film thickness, despite being immersed for only 10 s, which suggests that the rate of crystallization of a-$PbI_xBr_{2-x}$ was much quicker than the rate of formation of the perovskite structure. The diffusion of $FAI_xBr_{1-x}$ into the voids and/or grain boundaries resulted in the simultaneous formation of large amounts of free volume and an increase in the film thickness, resulting in a decrease in the refractive index (**Figure 4a**). The best-fit $d$ (=$d_1 + d_2 + d_3$), $d_1$, and $f_{void}$ for the top layer are summarized as a function of $t_{im}$ in **Figure 5c**, together with those determined using a thickness profiler. The value of $d$ increased from 150 nm for a-$PbI_xBr_{2-x}$

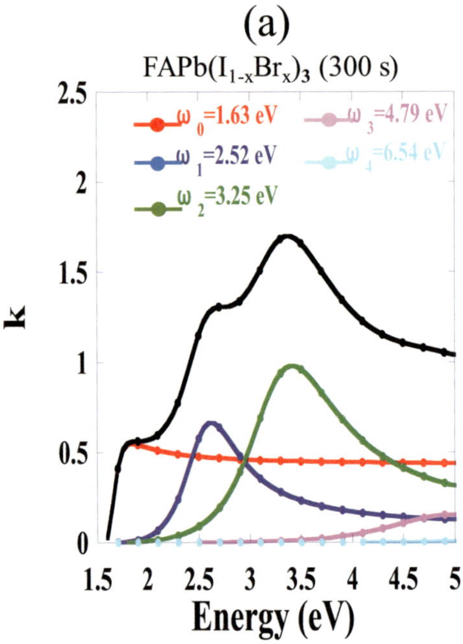


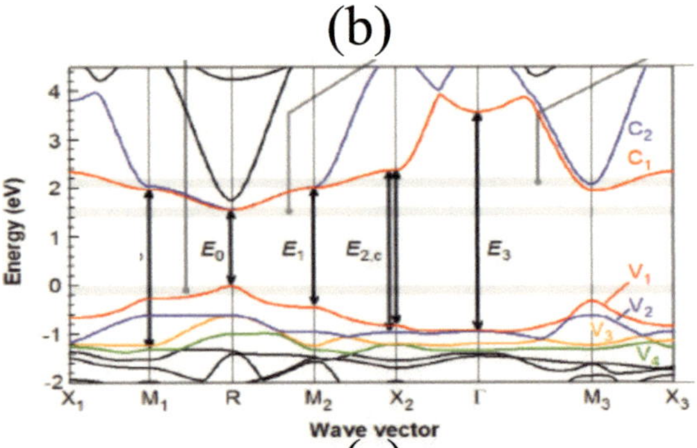


(c)

| Optical transition | $MAPbI_3$ | $FAPb(I_{1-x}Br_x)_3$ |
|---|---|---|
| $\omega_0$ | 1.61 (eV) | 1.63 (eV) |
| $\omega_1$ | 2.53 (eV) | 2.52 (eV) |
| $\omega_2$ | 3.24 (eV) | 3.25 (eV) |
| $\omega_3$ | 4.4 (eV) | 4.79 (eV) |
| | 7.29 (eV) | 6.54 (eV) |

**Figure 5.** (a) $k$ spectrum for the $FAPbI_xBr_{3-x}$ perovskite immersed in a solution of $FAI_xBr_{1-x}$ in IPA solvent for 600 s, including each optical transition component obtained by the spectral-fitting procedure. (b) Band structure of the $MAPbI_3$ perovskite determined from DFT calculations [26]. (c) Comparison of the optical transition components between $MAPbI_3$ [26] and $FAPbI_xBr_{3-x}$ perovskite.

to 350 nm for the $FAPbI_xBr_{3-x}$ perovskites. The volume fraction of voids and the thickness of $f_{void}$ and $d_1$ corresponding to the top layer increased significantly during the initial 10–60 s of immersion, after which they increased only gradually with $t_{im}$. These results imply that the formation of the perovskite structure proceeds with the formation of free space within the $PbI_xBr_{2-x}$ network due to the insertion of the organic FA cation.

**Figure 5a** shows the $n$ and $k$ spectra for the $FAPbI_xBr_{3-x}$ perovskite bulk component after immersion for 600 s, including each of the optical transition components determined by the deconvolution method. The electronic structure of $MAPbI_3$ determined from DFT calculations was reported by Shirayama et al. [26] (**Figure 5b**). The comparison between optical transition energies for $\omega_0$, $\omega_1$, $\omega_2$, and $\omega_3$ of $MAPbI_3$ and $FAPbI_xBr_{3-x}$ is summarized in **Figure 5c** [26, 27]. Similar to the band structure of $MAPbI_3$, four distinct optical transitions in $FAPb(I_xBr_{1-x})_3$ were observed at 1.63, 2.52, 3.25, and 4.79 eV, which may be attributed to direct semiconductor-type transitions at the R, M, X, and Γ points in the pseudo-cubic Brillouin zone, respectively. According to Shirayama et al., a sharp optical transition at 3.25 eV in $PbI_xBr_{1-x}$ is probably excitonic, and the transition energy of this peak is close to that of the peak for $MAPbI_3$. The absorption peak at 3.1 eV in a-$PbI_xBr_{2-x}$ is relatively sharp, while the corresponding peak was broad for $PbI_xBr_{2-x}$ after immersion for 10 s. Two distinct optical transitions at 1.77 and 2.5 eV were observed for the film immersed for 60 s. Here, the absorption peak at 2.5 eV is probably attributed to a layered structure of $PbI_xBr_{2-x}$ analogous to that of $PbI_2$ reported elsewhere [28].

The onset of the optical transition ($\omega_0$) was shifted to 1.75 eV and was almost independent of the immersion time over 300 s.

**Figure 6** shows the changes of the optical transition energies, $\omega_0$, $\omega_1$, $\omega_2$, and $\omega_3$ with $t_{im}$. The optical transition energies $\omega_0$, $\omega_1$, and $\omega_2$ shift to the infrared region and their amplitude increases when $t_{im}$ was increased up to 600 s. Among the four optical transitions, the lower-energy shift of the $\omega_2$ component from 3.85 to 3.4 eV was prominent and attributed to the 6*s*-6*p* transition of $Pb^{2+}$. Hirasawa et al. reported the exciton features in zero- (0D), two- (2D), and three-dimensional (3D) networks of $[PbI_6]^{4-}$ octahedra in $MA_4PbI_6 2H_2O$, $(C_{10}H_{21}NH_3)_2PbI_2$, and $MAPbI_3$ from reflection spectra measured at 4 K [29]. The energy of the lowest exciton decreases significantly as the dimension is increased and was 3.4, 2.55, and 1.633 eV for the 0D, 2D, and 3D network compounds, respectively. These results make an analogy that the process for the formation of the $FAPbI_xBr_{3-x}$ perovskite from a-$PbI_xBr_{2-x}$ is attributed to a decrease in the dimension of the crystal structure based on $[Pb(IBr)_6]^{4-}$.

### 3.3. Effect of *x* on the fine structure of $FAPbI_xBr_{3-x}$ perovskite films

**Figure 7** shows measured $I_s$ (=sin2ψsinΔ) and $I_c$ (=sin2ψcosΔ) spectra for $FAPbI_xBr_{3-x}$ perovskite films with two different $PbI_2$:$PbBr_2$ molar ratios of 3:1 and 3:2, including the best-fitted calculated spectra using the optical model shown on the right. The *n* of an I:Br = 3:2 film was larger than that of 3:1 film with increased film thickness and $f_{void}$. Thus, increased Br incorporation into $FAPbI_3$ promoted the densification of perovskite network, resulting in the increased film thickness and $f_{void}$, as far as a two-step method with immersion to FAIBr/IPA solution was

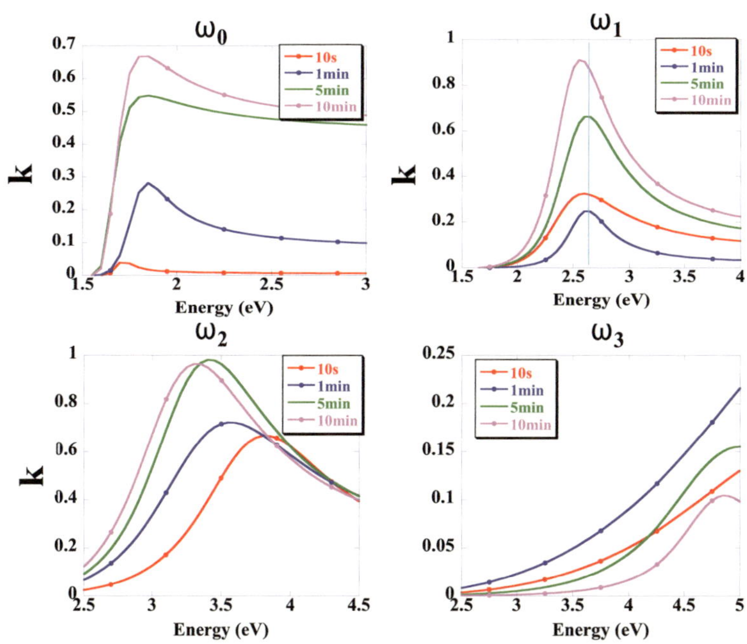


**Figure 6.** Changes of optical transition for $\omega_0$, $\omega_1$, $\omega_2$, and $\omega_3$ of $PbI_xBr_{2-x}$ immersed in $FAI_xBr_{1-x}$/IPA solution at various $t_{im}$. Each optical transition component was determined by the spectral-fitting procedure.

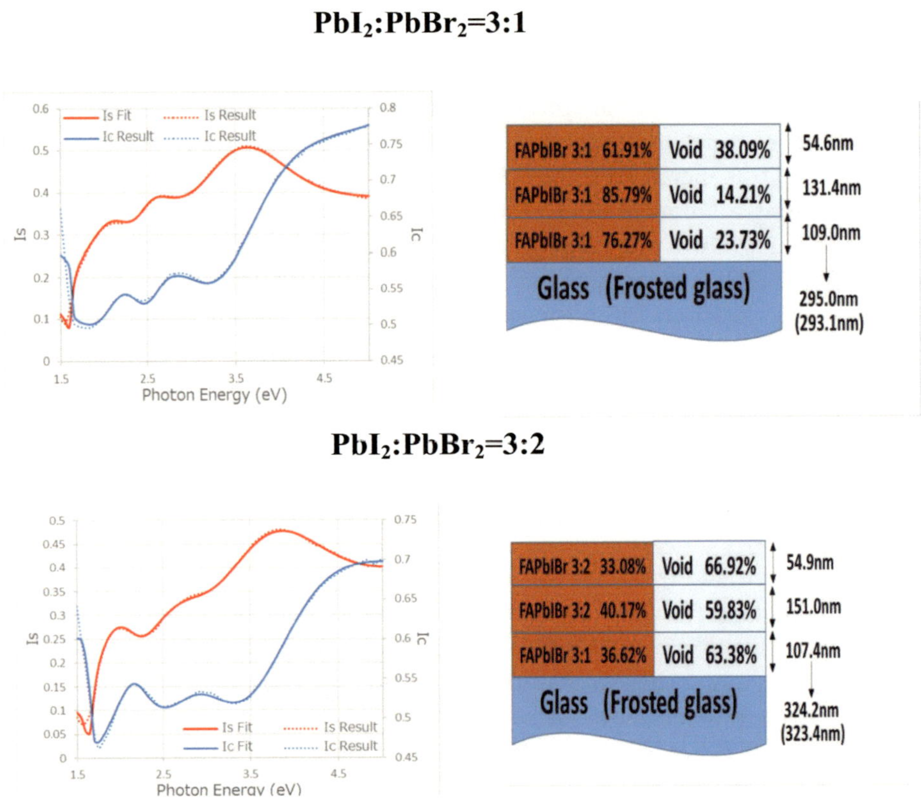


**Figure 7.** Measured $I_s$ and $I_c$ spectra of $FAPbI_xBr_{3-x}$ perovskite films with $PbI_2$:$PbBr_2$ = 3:1 and 3:2 molar ratio including the best-fitted calculated spectra. Corresponding best-fitted optical model.

used. In fact, XRD pattern and PL revealed that lattice parameter decreased with higher-edge emission peak energy when Br composition was increased. SE also showed an increase of the refractive index for $FAPbI_xBr_{3-x}$ perovskite phase together with an increase of the film thickness and volume fraction of void. In fact, RMS increased from 39 nm for 3:1 to 46 nm for 3:2 films in the 20 × 20-μm$^2$ area. These findings imply that Br incorporation into $FAPbI_xBr_{3-x}$ perovskite films promotes the densification of perovskite network, resulting in the increase in the free volume as a void and film thickness. In addition, prominent sub-gap absorption tail in the 3:2 film compared to that of 3:1 film suggests that the defect originates from the increased grain boundary of $FAPbI_xBr_{3-x}$ (3:2) perovskite phase for larger Br composition ratio.

### 3.4. Role of IPA solvent in the synthesis of $FAPbI_xBr_{3-x}$ perovskites

**Figure 8a** and **b** shows XRD pattern and FTIR spectra of a-$PbI_xBr_{2-x}$ film before and after immersion into IPA solvent alone for 10 s. The inset of **Figure 8a** also shows AFM images of the corresponding $PbI_xBr_{2-x}$ films. Despite immersion for only 10 s, the peak at $2\theta = 12.7°$ attributed to the crystalline $PbI_xBr_{2-x}$ (1 0 0) diffraction was observed, which suggests that the crystallization

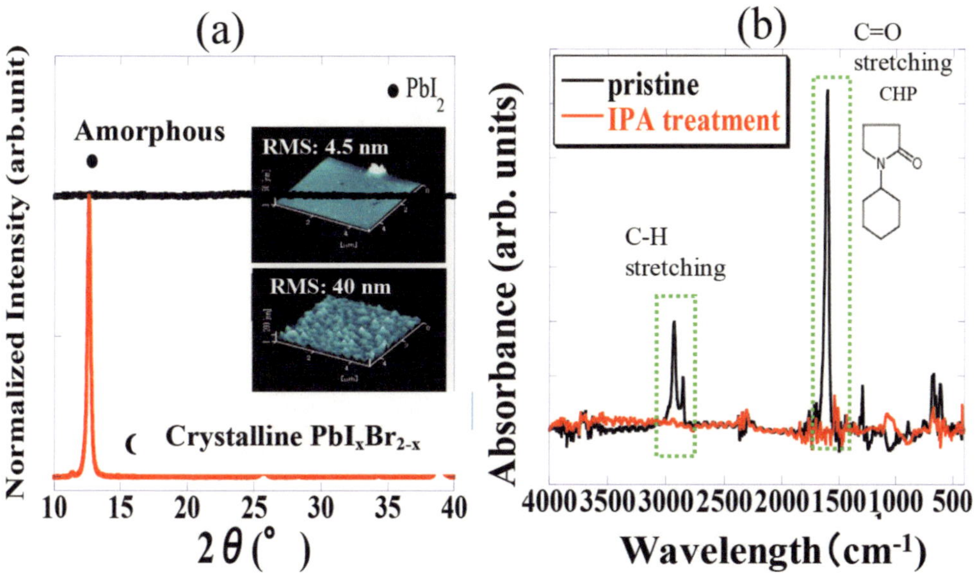


**Figure 8.** (a) XRD pattern and (b) FTIR spectra of a-PbI$_x$Br$_{2-x}$ film before and after immersion into IPA solvent alone for 10 s. The inset of **Figure 8a** shows the AFM image of corresponding PbI$_x$Br$_{2-x}$ before and after immersion into IPA solvent alone without FAI$_x$Br$_{1-x}$.

of the film can be attributed mainly to immersion in the IPA solvent. The FTIR spectra also revealed that the removal of the C=O-bending mode and C—H-stretching mode attributed to residual CHP in a-PbI$_x$Br$_{2-x}$ was promoted by the IPA solvent. Thus, film crystallization of a-PbI$_x$Br$_{2-x}$ was promoted by the removal of residual CHP in a-PbI$_x$Br$_{2-x}$ and was accompanied by the formation of mesoporous crystallized PbI$_x$Br$_{2-x}$ with large amounts of voids and/or grain boundaries.

**Figure 9** shows $n$ and $k$ spectra for a-PbI$_x$Br$_{2-x}$ before and after immersion into IPA solvent alone for 10 s. The best-fit optical models are shown in the upper panels. A prominent feature was that the $n$ value decreased markedly with a red shift of the absorption edge onset corresponding to film crystallization with an increase in $f_{void}$. The significant decrease in $n$ is due to crystallization and the destruction of the mesoporous in a-PbI$_x$Br$_{2-x}$. These results imply that IPA had a significant effect to promote crystallization of the a-PbI$_x$Br$_{2-x}$ film with an associated decrease in the bandgap due to a phase transition from 0D to 3D $[Pb(I_xBr_{1-x})_6]^{4-}$ octahedral clusters. These factors are the main contributors to the formation of the FAPbI$_x$Br$_{3-x}$ perovskite network. Notable spectral features were the appearance of broad and sharp absorptions at 2.2 (probably due to defects/disorder) and 2.48 eV ($A_4^+ \rightarrow A_4$) assigned as the first exciton region in 2D PbI$_x$Br$_{2-x}$, in addition to those at 3.1, 3.5, and 3.9 eV due to film crystallization. In addition, no significant change in the film thickness was observed after immersion into the IPA solvent, but only the formation of voids, which suggests that immersion in IPA did not contribute to an increase of the film thickness.

From these results, the growth mechanism for the formation of FAPbI$_x$Br$_{3-x}$ perovskites from an a-PbI$_x$Br$_{2-x}$ layer and FAI$_x$Br$_{1-x}$/IPA solution is considered to be as follows (**Figure 10**) [30].

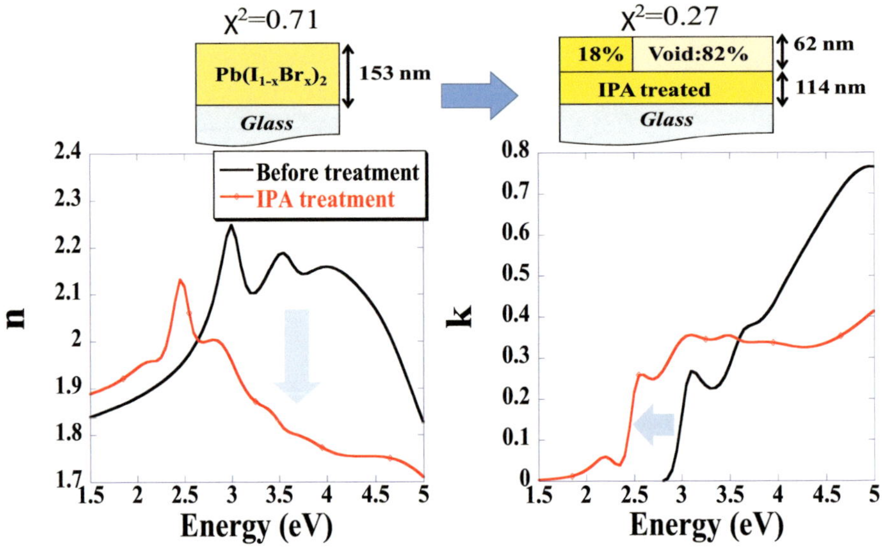


**Figure 9.** Bulk $n$ and $k$ spectra component for $PbI_xBr_{2-x}$ films on glass before (black) and after (red) immersion into IPA solvent for 10 s. The best-fit optical model and $\chi^2$ value are shown on the top.

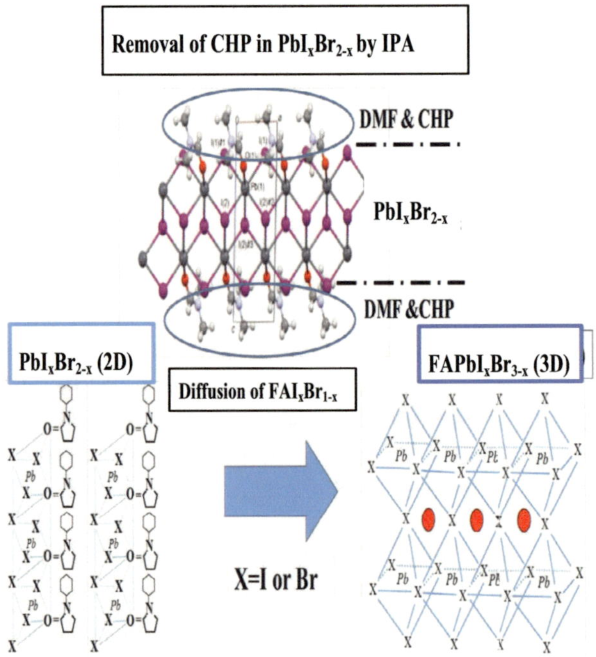


**Figure 10.** Schematic of the formation mechanism of $FAPbI_xBr_{3-x}$ perovskites from a-$PbI_xBr_{2-x}$ layer and a solution of $FAI_xBr_{1-x}$ in IPA solvent. The formation of the perovskite structure is accompanied by the removal of coordinated CHP in the a-$PbI_xBr_{2-x}$ network during immersion in the IPA solvent.

http://dx.doi.org/10.5772/intechopen.70281

$FAPbI_xBr_{3-x}$ perovskites include an octahedral $PbX_6$ (X = I, Br) complex with coordinated CHP solvent molecules. When a-$PbI_xBr_{2-x}$ including —CHP groups is immersed in the IPA solvent, C═O and C—H groups in coordinated CHP are preferentially removed by IPA solvent, which results in the promotion of a-$PbI_xBr_{2-x}$ film crystallization and the formation of a large amount of voids and/or grain boundaries in the network. The diffusion of $FAI_xBr_{1-x}$ into the voids/grain boundaries of the crystalline $PbI_xBr_{2-x}$ phase is simultaneously promoted and leads to the formation of the $FAPbI_xBr_{3-x}$ perovskite structure with an increase in the film thickness. In addition, no XRD diffraction peaks attributed to the perovskite structure were observed for vacuum-evaporated $FAI_xBr_{1-x}$ as a precursor layer without a polar solvent such as IPA and CHP on a-$PbI_xBr_{2-x}$ films, which implies that the IPA solvent promotes the growth of the perovskite structure.

## 4. Conclusions

1. The effect of Br incorporation into $FAPbI_3$ perovskite films is to promote the densification of perovskite network, resulting in the increase in the free volume as a void and film thickness. In addition, prominent sub-gap absorption tail in the 3:2 film compared to that of 3:1 film suggests that the defect originates from the increased grain boundary of perovskite phase for larger Br compositional ratio.

2. The role of IPA solvent in the synthesis of $FAPbI_xBr_{3-x}$ perovskites from an a-$PbI_xBr_{2-x}$ thin layer and a solution of $FAI_xBr_{1-x}$ in IPA solvent was investigated using SE combined with characterization by XRD and FTIR analysis. IPA played a significant role to promote crystallization of the a-$PbI_xBr_{2-x}$ film through the removal of coordinated CPH from the a-$PbI_xBr_{2-x}$ network, accompanied by the formation of grain boundaries, voids, and surface roughness. The diffusion of $FAI_xBr_{1-x}$ into the voids and/or grain boundaries resulted in the simultaneous formation of large amounts of free volume and an increase in the film thickness, resulting in a decrease in the refractive index. The red shift of absorption edge from 3.4 to 1.73 eV is attributed to a phase transition from 0D to 3D $[PbX_6]^{4-}$ octahedral clusters. These factors are the main contributors to the formation of the $FAPbI_xBr_{3-x}$ perovskite network.

## Acknowledgements

The author thanks Dr. Yohei Numata and Professor Tsutomu Miyasaka of Toin Yokohama University for fabrication of the perovskite thin film solar cells.

## Author details

Hajime Shirai

Address all correspondence to: shirai@fms.saitama-u.ac.jp

Graduate School of Science and Engineering, Saitama University, Saitama, Japan

## References


[1] Kojima A, Terashima K, Miyasaka T, Shirai Y. Novel halide compounds (2). In: 210th ECS Meeting. Pennington, NJ: The Electrochemical Society; 2006. p. 397

[2] Ishii A, Jena AK, Miyasaka T. Fully crystalline perovskite-perylene hybrid photovoltaic cell capable pf 1.2 V output with a minimized voltage loss. APL Materials. 2014;**2**:091102

[3] Green MA, Ho-Ballie A, Snaith HJ. The emergence of perovskite solar cells. Nature Photonics. 2014;**8**:506-514

[4] Kim HS, Lee CR, Im J-H, Lee KB, Moehl T, Marchiro A, Moon SJ, Humphry-Baker R, Yum J-H, Moser JE, Grätzel M, Park NG. Lead iodine perovskite sensitized all-solid state sub-micron thin film mesoscopic solar cell with efficiency exceeding 9%. Scientific Reports. 2012;**2**:1-7

[5] Eperon GE, Stranks SD, Menelaou C, Johnson MB, Herz LM, Snaith HJ. Formamidinium lead trihalide: A. Broadly tunable perovskite for efficient planer heterojunction solar cells. Energy & Environmental Society. 2014;**7**:982-988

[6] De Wolf S, Holovsky J, Moon SJ, Löper P, Niesen B, Ledlinsky M, Haug FJ, Yum J-H, Ballif C. Organometallic halide perovskites: Sharp optical absorption edge and its relation to photovoltaic performance. Journal of Physical Chemistry Letters. 2014;**5**:1035-1039

[7] Snaith HJ. Perovskites: The emergence of a new era for low-cost, high-efficiency solar cells. Journal of Physical Chemistry Letters. 2013;**4**:3623-3630

[8] Topic M, Geisthardt RM, Sites JR. Performance limits and status of single junction solar cells with emphasis on sigs on CIGS. IEEE Journal of Photovoltaics. 2014;**5**:1-6

[9] Jeon NJ, Noh JH, Yang WS, Kim YC, Ryu S, Seo J, Seok SI. Compositional engineering of perovskite materials for high-performance solar cells. Nature. 2015;**517**:476-480

[10] Kojima A, Teshima K, Shirai Y, Miyasaka T. Organometal halide perovskites as visible-light sensitizers for photovoltaic cells. Journal of the American Chemical Society. 2009;**131**:6050-6051

[11] Burschka J, Pellet N, Moon S-J, Humphry-Baker R, Gao P, Nazeeruddin MK, Grätzel M. Sequential deposition as a route to high-performance perovskite-sensitized solar cells. Nature. 2013;**499**:316-320

[12] Im J-H, Jang I-H, Pellet N, Grätzel M, Park N-G. Growth of $CH_3NH_3PbI_3$ cuboids with controlled size for high-efficiency perovskite solar cells. Nature Nanotechnology. 2014;**9**:927-932

[13] Liu M, Johnston MB, Snaith HJ. Efficient planar heterojunction perovskite solar cells by vapor deposition. Nature. 2013;**501**:395-398

[14] Jeon NJ, Noh JH, Kim YC, Yang WS, Ryu S, Seok SI. Solvent engineering for high-performance inorganic–organic hybrid perovskite solar cells. Nature Materials. 2014;**13**:897-903

[15] Xiao M, Huang F, Huang W, Dkhissi Y, Zhu Y, Etheridge J, Gray-Weale A, Bach U, Cheng Y-B, Spiccia L. A fast deposition-crystallization procedure for highly efficient lead iodide perovskite thin-film solar cells. Angewandte Chemie International Edition. 2014;**53**:9898-9903

[16] Song J, Zheng E, Bian J, Wang X-F, Tian W, Sanehira Y, Miyasaka T. Low-temperature $SnO_2$-based electron selective contact for efficient and stable perovskite solar cells. Journal of Materials Chemistry. 2015;**A3**:10837-10844

[17] Yang L, Wang J, Leung WW-F. Lead iodide thin film crystallization control for high-performance and stable solution-processed perovskite solar cells. ACS Applied Materials & Interfaces. 2015;**7**:14614-14619

[18] Jo Y, Oh KS, Kim M, Kim K-H, Lee H, Lee C-W, Kim DS. High performance of planar perovskite solar cells produced from $PbI_2$(DMSO) and $PbI_2$(NMP) complexes by intramolecular exchange. Advanced Materials Interfaces. 2016;**3**:1-7

[19] Miyadera T, Shibata T, Koganezawa T, Murakami TN, Sugita T, Tanigaki N, Chikamatsu M. Crystallization dynamics of organolead halide perovskite by real-time X-ray diffraction. Nano Letters. 2015;**15**:5630-5634

[20] Wu Y, Islama A, Yang X, Qin C, Liu J, Zhang K, Penga W, Han L. Retarding the crystallization of $PbI_2$ for highly reproducible planar-structured perovskite solar cells *via* sequential deposition. Energy and Environment Science. 2014;**7**:2934-2938

[21] Jeon YJ, Lee S, Kang R, Kim JE, Yeo JS, Lee SH, Kim SS, Yun JM, Kim DY. Planar heterojunction perovskite solar cells with superior reproducibility. Scientific Reports. 2014;**6953**:1-7

[22] Forouhi A, Bloomer I. Optical dispersion relations for amorphous semiconductors and amorphous dielectrics. Physical Review B. 1986;**34**:7018-7026

[23] Jobin Yvon J. New amorphous dispersion formula. (accessed Oct29, 2015). Available from: http://www. Horiba. Com/fleadmi/New_amorphous_dispersion_formula pdf

[24] Aspnes DE. Optical properties of thin films. Thin Solid Films. 1982;**89**:249-262

[25] Schliiter ICh, Schliiter M. Electronic structure and optical properties of $PbI_2$. Physical Review B. 1974;**9**:1652-1665

[26] Shirayama M, Kadowaki H, Miyadera T, Sugita T, Kato M, Murata D, Hara S, Murakami T, Chikamatsu M, Fujiwara H. Optical transitions in hybrid perovskite solar cells. Physical Review Applied. 2016;**5**:01402

[27] Lopper P, Stuckellger M, Niesen B, Werner J, Filpie M, Moon SJ, Yum JH, Topic M, Wolf D, Ballif C. Complex refractive index spectra of $CH_3NH_3PbI_3$ perovskite thin films determined by spectroscopic ellipsometry and spectrophotometry. Physical Chemistry Letters. 2015;**6**:66-71

[28] Ishihara T, Takahashi J, Goto T. Optical properties due to electronic transitions in two-dimensional semiconductors $(C_nH_{2n}+1NH_3)_2PbI_4$. Physical Review. 1990;**B42**:11099-11107

[29] Hirasawa M, Ishihara T, Goto T. Exciton features in 0-, 2-, 3-dimensional networks of $[PbI_6]^{4-}$ octahedra. Journal of the Physical Society of Japan. 1994;**63**:3870-3879

[30] Yamanaka T, Masumori K, Ishikawa R, Shirai H. Role of isopropyl alcohol (IPA) in the synthesis of perovskite thin films by two-step method. Journal of Physical Chemistry. 2016;**120**(44):25371-25377

Chapter 6

# Photoelastic Modulated Imaging Ellipsometry

Chien-Yuan Han, Yu-Faye Chao and Hsiu-Ming Tsai

Additional information is available at the end of the chapter

http://dx.doi.org/10.5772/intechopen.70254

**Abstract**

Photoelastic modulator (PEM)-based ellipsometry employed either lock-in amplifiers or the Fourier analysis technique to obtain the ellipsometric parameters almost in real-time that makes the system with a feature of fast measurement speed, higher stability, and sensitivity at small retardations. Since the PEM modulation frequency is too high to compare it with the exposure time of the camera, photoelastic modulator–based approach is not applicable for a two-dimensional ellipsometric measurement. Here, we represent a novel technique that coordinates with the light pulses and PEM modulation that can freeze the time-varied signals. Thus, two-dimensional ellipsometric parameters can be obtained within few seconds. In addition to ellipsometric measurement, this approach also can be extended to other imaging polarimetry measurements, such as Stokes parameters and Mueller matrix. Moreover, since the chromatic dispersion of birefringence was also a significant issue in the polarization modulation systems, we proposed an equivalent phase retardation technique to deal with this issue. This technique was confirmed by a dual wavelength measurement result without changing the optical configuration of the system. The concept and the theory of this system were indicated in the preceding section, and the passage below described some calibration issues for the photoelastic modulator. Some measurement results were revealed in the final part of this chapter.

**Keywords:** photoelastic modulator, imaging polarimetry, ellipsometric parameters, Stokes parameter, Mueller matrix, dual wavelength

## 1. Introduction

Among other optical techniques, ellipsometry is one of the most powerful tools for characterizing optical properties, including determining film thicknesses and refractive indices with a high degree of accuracy. Currently, common applications of ellipsometry include measuring thin films for solar cells, optical coatings, microelectronics, and biosensing applications [1–3]. However, most ellipsometric measurements are based on a single point and use a single-wavelength or spectroscopic approach [4, 5]. As the size of many electronic devices becomes smaller, the uniformity of the thin-film thickness, a high degree of resolution, and a large field of view become more desirable for industrial applications. Therefore, ellipsometry with a spatially resolved capability to assess two-dimensional morphologies of a surface is a natural evolution of ellipsometric measurement techniques that extend the single spot ellipsometric measurement to a tool to visualize and analyze microscopic images for thin films [6, 7]. Thus far, the commercial imaging ellipsometer usually operated on the principle of the classical null technique; instruments used are typically equipped with stepping motors to change the azimuth angle of the polarizer, compensator, or analyzer; and they use charge-coupled device (CCD) or complementary metal-oxide-semiconductor (CMOS) detectors to take a sequence of images in order to gather enough information to calculate all null positions [6]. This approach is relatively slow and is limited by the mechanical rotation speed, and the modulation frequency usually falls within a noise range of other mechanical devices; this impedes data acquisition and, eventually, system stability. In addition, for a sample with inhomogeneous surface characteristics, the measurement process may need to collect many more images to determine the null positions of each measurement point; this makes the measurements relatively cumbersome and impractical for industrial applications.

Another popular approach for imaging ellipsometry was using photometric measurement technique, which few intensity images captured at various angles of polarization elements to deduce ellipsometric distributions of the sample [8–10]. However, this approach requires the rotation of polarization elements and suffers the issue of beam wander during the rotation, possibly resulting in mismatching the interest point in a sample and a recording pixel associated with different images. The above issues make the measurement troublesome especially for spatially nonuniform objects or imperfectly uniform incident beam. Compared with null ellipsometry, photometric ellipsometry is faster, and the measurement quality is improved. Phase modulated apparatus is also involved in the ellipsometric measurement. The ellipsometer based on the use of a photoelastic modulator (PEM) is the most prevalent configuration; it has a typical modulation frequency of 50 kHz with no moving parts [11–13]. For single spot measurement, one can employ either lock-in amplifiers or the Fourier analysis technique to obtain the ellipsometric parameters in near real-time, but this approach is not applicable for a two-dimensional measurement, because the modulation frequency of the PEM is too high to compare it with the exposure times of the CCD camera. This deficiency was overcome by replacing the light source with an ultra-stable short pulse, known as the stroboscopic illumination technique, which was synchronized with the PEM modulation to freeze the intensity signal at specific times in the modulation cycles [14, 15]. As a result, the ellipsometric images can be obtained in seconds by sequentially taking four images with single wavelength source. Since Stokes parameters were related to the ellipsometric parameters by their definitions, Stokes parameters also can be measured through the use of the same approach [16]. In order to remove the chromatism limits of this system, we adopted an equivalent phase retardation technique that used dual wavelength ellipsometric measurement to extend the imaging ellipsometry technique without any adjustment of the photoelastic modulator or the optical configuration. Moreover, the image acquisition time for one set of ellipsometric parameters for dual-wavelength measurement remained virtually unchanged [17]. This technique was also applied in Mueller matrix imaging system; we introduced a hybrid phase modulation technique to evaluate the optical polarization characteristics of the specimens [18]. In this chapter, we introduced the principle of the system, explained how this concept can be used in imaging ellipsometric measurement, and demonstrated dynamic measurement results. As PEM is the crucial component in this system, some calibration processes are also discussed in this chapter.

## 2. Theory and optical configuration of the photoelastic modulated imaging ellipsometry

Ellipsometry measures the changes of a polarized light that is reflected from the sample surface; such changes can be used to deduce the optical parameters of the sample. The ellipsometric parameters, Ψ and Δ, are defined as:

$$\tan \Psi e^{i\Delta} = \frac{r_p}{r_s} \tag{1}$$

where $r_p$ and $r_s$ are the complex Fresnel reflection coefficients for polarized light that is parallel and perpendicular to the plane of incidence, respectively [4]. The compensator was replaced by a PEM in the polarizer-compensator-sample-analyzer setup, as shown in **Figure 1**. The output polarization state can be represented by the operation of their corresponding Mueller matrices; i.e., the polarization state can be expressed as:

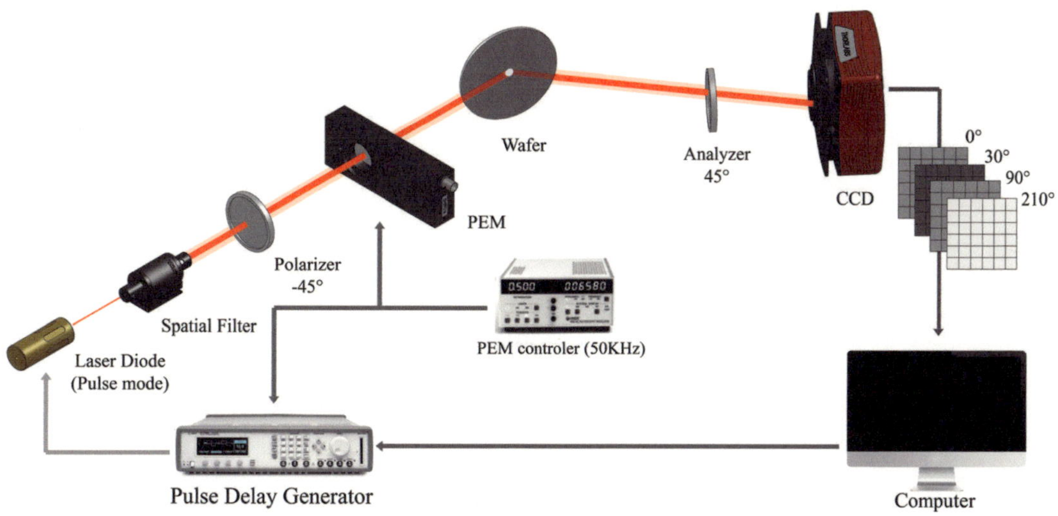


**Figure 1.** Experimental setup of the photoelastic modulated imaging ellipsometry by stroboscopic illumination technique.

$$S_f = M_A(A) \cdot R_{SAM}(\Psi, \Delta) \cdot M_{PEM}(\theta, \Delta_P) \cdot S_P \tag{2}$$

where $S_f$ and $S_P$ are Stokes vectors of the output polarization state and the incident linearly polarized light at the azimuth angle of P, respectively. Moreover, $M_{PEM}(\theta,\Delta_P)$, $R_{SAM}(\Psi, \Delta)$, and $M_A(A)$ represent the Mueller matrix of the PEM, the sample, and the analyzer, respectively. In this configuration, the optic axis of the PEM is at 0° with respect to the incident plane. When P = −45° and A = 45°, the reflected intensity can be found to be:

$$I(t) = \frac{I_0}{2}[1 + \sin 2\Psi \cos(\Delta - \Delta_P) \tag{3}$$

where $I_0$ is the normalized output intensity, and $\Delta_P$ is the phase retardation of the PEM, which is modulated as $\delta_0 \sin\omega t$. If one set of the amplitude of modulation $\delta_0$ equals $\pi$, the temporal intensity behavior can be formulated as:

$$I(t) = \frac{I_0}{2}[1 + \sin 2\Psi \cos(\Delta - \pi \sin \omega t)] \tag{4}$$

When the temporal phase angles $\omega t$ in Eq. (4) are 0 and 90°, the corresponding intensities can be expressed as:

$$I_{0^\circ} = \frac{I_0}{2}[1 + \sin 2\Psi \cos \Delta] \tag{5}$$

and

$$I_{90^\circ} = \frac{I_0}{2}[1 - \sin 2\Psi \cos \Delta] \tag{6}$$

respectively. It is easy to prove that:

$$\sin 2\Psi \cos \Delta = \frac{I_{0^\circ} - I_{90^\circ}}{I_{0^\circ} + I_{90^\circ}} = I' \tag{7}$$

Using the similar process for $\omega t$ at 30 and 210°, one can obtain:

$$\sin 2\Psi \sin \Delta = \frac{I_{30^\circ} - I_{210^\circ}}{I_{30^\circ} + I_{210^\circ}} = I'' \tag{8}$$

Thus, the ellipsometric parameters can be obtained by measuring the intensity at above four temporal phases, as follows:

$$\Delta = \tan^{-1}\left(\frac{I''}{I'}\right) \tag{9}$$

$$\Psi = \frac{1}{2}\sin^{-1}\left(\sqrt{I'^2 + I''^2}\right) \tag{10}$$

The time-varying signal can be frozen at these temporal phases by illuminating the objects with short synchronized light pulses.

### 2.1. Interpretation of stroboscopic illumination applied in PEM imaging system

For a conventional phase modulated ellipsometer, photoelastic modulator is the most prevalent modulator installed in the ellipsometric system. The photoelastic modulator oscillates at its resonance frequency (typically around 50 kHz), and the response time of the phase modulated ellipsometer can reach as short as 1 ms/point to achieve the requirement for real-time monitoring and dynamic studies [19–21]. Although the ellipsometer using photoelastic modulator to modulate or examine polarization states has the advantages of being very fast, having no moving elements for acquiring signals, the high frequency of modulation was also an issue for the two-dimensional measurement due to the fact that the modulation frequency is much higher than the image sensor frame rate. To overcome this issue, two approaches are developed in recent days. One approach was using two or four PEMs and field-programmable gate array (FPGA)-assisted sequential time gating approach. In that configuration, four PEMs are set at different azimuths, and their modulation frequencies also have to be different from each other. The frequency drift of PEMs and image recovery were taken more effort on calibration and measurement [22–25]. We adopted a simple approach by just changing the manner of light illumination from continuous mode to pulse mode, which coordinates with the reference signal of the PEM modulator, known as the stroboscopic illumination technique [14–18]. Stroboscopic effect is a result of temporal aliasing that occurs when continuous motion is represented by a series of short samples. If the motion is circular or repeating, such as a spinning wheel or a vibrating membrane, and the frequency of light pulses and wheel speed or membrane oscillation are the same, the wheel or the membrane will appear stopped. In our system, PEM modulator functioned as a resonant device and operated at a fixed frequency about 50 kHz; therefore, while the light pulses coordinate with the resonant frequency of the PEM, the phase retardation of the modulator can be fixed at a specific value, rather than a continuous variation as a function of time. The synchronization process between the light pulses and PEM modulator was through the square wave reference signal, as the extra-trigger, of the PEM controller to initiate light pulses whose width was 110 nm (~2° phase change of modulator) from diode lasers. Another issue to be considered was how to shift the phase of the modulated optical signal that can generate different polarizations of the outgoing light. This feature can be achieved by the digital delay function of the pulse generator that provides defined pulses at four specific intervals, which are 0, 30, 90, and 210°, as shown in **Figure 2**. The $\omega t = 0°$ pulse, the beginning of a modulation cycle, is generated by the pulse generator, and then the time shift delays of the pulses are sequentially set on the basis of temporal phase angle $\omega t$, i.e. 30, 90, and 210°, of the PEM modulation. After setting the light pulses at the proper triggering and delay outputs, the output polarizations can appear to be frozen by use of the stroboscopic illumination technique. For the image acquisition, the exposure time $\Delta t$ of the CCD camera was at the range of decisecond to several seconds depending on the intensity of the diode laser and the sensitivity of the camera. For maintaining the intensity in the linear range, four specific images at different temporal phase angles were obtained to deduce the two-dimensional ellipsometric parameters.

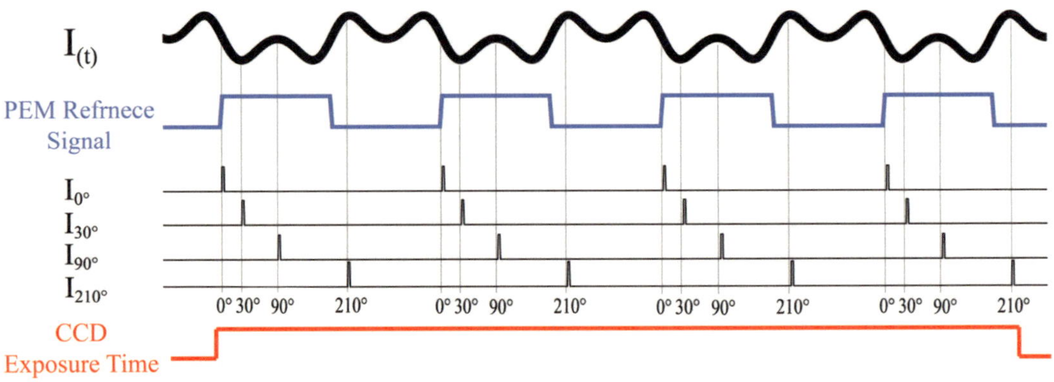


**Figure 2.** The principle of image acquisition: each intensity ($I_{0°}$, $I_{30°}$, $I_{90°}$, $I_{210°}$) was obtained by the accumulation of N short pulses of the modulated signal I(t) at a specific temporal phase angle at the fixed exposure time of the camera. The four intensities were acquired in sequence by the synchronized ultrastable short-pulse illumination. The ellipsometric images are calculated by these four intensities.

## 2.2. Issues in calibration of the PEM imaging system

All ellipsometric methods require azimuthal alignment of the polarizers, retarders, and phase modulators, with respect to the plane of incidence. If this alignment is not accurate, the systematic error appears in the ellipsometric measurements. Overlooking the azimuthal alignment of the polarizer and analyzer, there are some alignment issues that have to be addressed in a PEM-based ellipsometer and that are discussed below:

### *2.2.1. Azimuth angle calibration of the photoelastic modulator*

It is essential to align the azimuths of the optical components in the ellipsometer for accurate measurement because any improper azimuth setting in the system can cause significant errors. The null method, locating the minimum intensity, is a typical azimuthal alignment technique in most ellipsometric systems. Since the minimum intensity must be determined precisely in the null method, a highly sensitive detection apparatus is required in those techniques. Instead of using the null method, we proposed an intensity ratio technique and separately aligned the azimuths of the polarizer and analyzer to the specimen surface in a polarizer-sample-analyzer (PSA) system [26, 27]. After precisely locating the incident plane in the PSA system, we then shift the attention to determine the strain axis of PEM to the incident plane. If the strain axis of PEM deviates from the incident plane by θ, and the impinging light is a +45° linear polarized light, the intensity can be reformulated from Eq. (5) as:

$$I_{(A)} = I_0\left[L\sin^2 A + M\cos^2 A\tan^2\Psi + N\sin A\cos A\right] \tag{11}$$

where

$$L = \frac{1}{2}[1 + \cos\Delta_P + (1 - \cos\Delta_P)(1 - \sin 4\theta)]$$

$$M = \frac{1}{2}[1 + \cos\Delta_P + (1 - \cos\Delta_P)(1 + \sin 4\theta)]$$

$$N = \frac{1}{2}[1 + \cos\Delta_P - (1 - \cos\Delta_P)\cos 4\theta]\tan\Psi\cos\Delta - \sin\Delta_P\cos 2\theta\tan\Psi\sin\Delta$$

When the A = 0 and 90°, the expression of intensity can be reduced and expressed as:

$$I_{(0^\circ)} = \frac{I_0}{2}\tan^2\Psi[2 + (1 - \cos\Delta_P)\sin 4\theta] \tag{12}$$

$$I_{(90^\circ)} = \frac{I_0}{2}[2 - (1 - \cos\Delta_P)\sin 4\theta] \tag{13}$$

If the phase modulation $\Delta_P$ is modulated as $\delta_0\cos\omega t$, then the intensity can be Fourier expanded by its harmonic function:

$$\cos\Delta_P = J_o(\delta_o) - 2J_2(\delta_o)\cos 2\omega t \tag{14}$$

By taking the zero-order Bessel function $J_0(\delta_0)$ at its zero point, i.e., $\delta_0 = 0.383\lambda$, we can simplify the DC component of its intensity as:

$$I_{(0^\circ)} = \frac{I_0}{2}\tan^2\Psi[2 + \sin 4\theta] \tag{15}$$

$$I_{(90^\circ)} = \frac{I_0}{2}[2 - \sin 4\theta] \tag{16}$$

From Eqs. (14) and (15), the azimuth deviation of the PEM can be obtained by the DC component of the intensity, which is taken at two azimuths of the PEM separated by 45° through the following relation:

$$\sin 4\theta_o = 2\frac{I_{dc}(0^\circ)_{\theta=\theta_0} - I_{dc}(0^\circ)_{\theta=\theta_0+45^\circ}}{I_{dc}(0^\circ)_{\theta=\theta_0} + I_{dc}(0^\circ)_{\theta=\theta_0+45^\circ}} \tag{17}$$

In addition to the azimuth determination, the ellipsometric parameter $\Psi$ can also be obtained by the same measurements as:

$$\tan^2\Psi = \frac{I_{dc}(0^\circ)_{\theta=\theta_o+45^\circ}}{I_{dc}(90^\circ)_{\theta=\theta_o}} \tag{18}$$

Eq. (16) is sufficiently general to analyze the error of the azimuth deviation. According to the intensity ratio of Eq. (16), one can easily prove that the deviation of azimuth $\delta\theta_0$ caused by those fluctuations is:

$$\delta\theta_o = \frac{\tan 4\theta_0}{4\tan^2\Psi}\frac{\delta I}{I} \tag{19}$$

*2.2.2. Amplitude modulation calibration of the photoelastic modulator*

Even if all the azimuths of the optical components can be aligned in a PEM ellipsometer at a fixed incident angle, the modulation amplitude of PEM still needs to be calibrated. The conventional technique for calibrating the modulation amplitude is to adjust the oscilloscope waveform of a half-wave modulation in a straightforward setup, and a multiple-paths method is used to amplify the modulation amplitude for higher resolution. However, this technique was not proceeded under reflection configuration and did not meet the requirement of in-situ calibration of the ellipsometer. We introduce a technique for the calibration of the modulation amplitude of PEM by a multiple harmonic intensity ratio (MHIR) technique whose setup was the same as for the ellipsometric measurement [28]. As a result, the modulation amplitudes of the PEM can be determined by using the intensity ratios of $I_{1f}/I_{3f}$ (odd ratio) and $I_{2f}/I_{4f}$ (even ratio) by the following:

$$\frac{I_{1f}}{I_{3f}} = \frac{J_1(\delta_0)}{J_3(\delta_0)}; \frac{I_{2f}}{I_{4f}} = \frac{J_2(\delta_0)}{J_4(\delta_0)} \tag{20}$$

It is clear that these ratios are independent of the azimuth position of the analyzer and the physical parameters of the examined sample. In this way, the optical characteristics of the PEM can be completely recognized.

### *2.2.3. Initial phase determination of the photoelastic modulator*

Compared with the conventional continuous wave modulation, our approach employed the pulse lights initiated at different phase angles. Therefore, the additional condition, initial phase angle of the pulse light, has to be checked before the measurement. In the previous section, we demonstrated that four temporal phase angles $\omega t$, 0, 30, 90, and 210°, were set to initiate pulse lights, triggered by the external square wave, from the PEM controller. However, we found reference zero of the square wave does not match with the initial phase of optical modulation signal, as shown in **Figure 3**, which means further determination of the phase shift of both signals is required. The determination process was carried out by an additional intensity

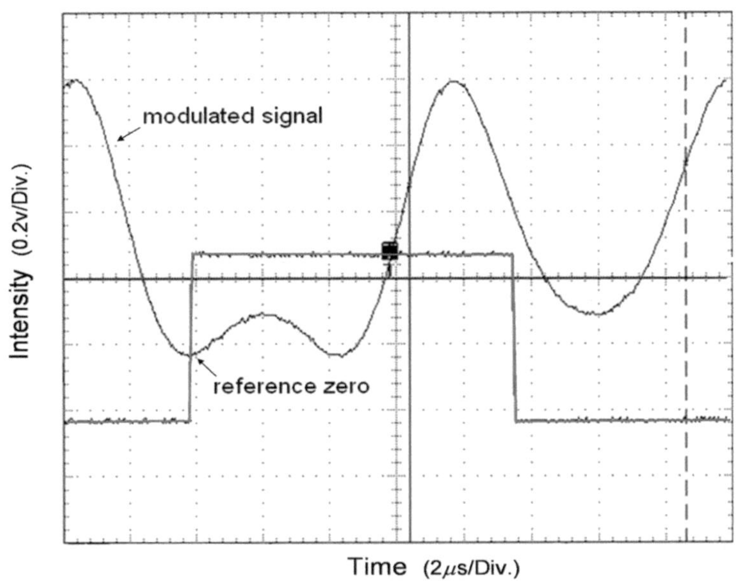


**Figure 3.** The temporal waveform of the Pt/Si thin film and the reference square wave provided by the PEM; the local minimum intensity does match with the reference zero.

measurement at ωt = 180°. If the phase shift between the square reference signal and modulated optical signal is x, one can determine x by the following equation:

$$\frac{I_{210^\circ} - I_{30^\circ}}{I_{180^\circ} - I_{0^\circ}} = \frac{\sin(\pi \sin(x))}{\sin(\pi \sin(x + \pi/6))} \tag{21}$$

This ratio can eliminate the effect of ellipsometric parameters and normalized intensity, so it is free from the material under investigation. As a result, the phase shift can be solved by the intensity measurements at ωt = 0, 30, 180, and 210° according to Eq. (20). The correction of phase shift x can also be achieved by the time shift delays of the pulses [15].

## 3. Two-dimensional measurement results for ellipsometric parameters, Stokes parameters, and Mueller matrix

### 3.1. PEM imaging system for the static ellipsometric measurement

An L-shaped $SiO_2$ layer with the thickness of 50 nm on a silicon substrate was set as the static sample to examine the feasibility of the stroboscopic illumination imaging ellipsometry. Before the examination, the light beam was expanded to cover the whole L-shaped pattern, and the deduced thickness profile is shown in **Figure 4**. One can observe the plateau of the thickness profile was about 52 nm, which was consistent with the thickness before etching. The inset of **Figure 4** shows the valley of the profile is 2 nm oxide layer after the etching process, and this result was ignored at the 2° static phase retardation of PEM.

### 3.2. PEM imaging system for a dynamic ellipsometric measurement

Besides the static measurement, an oil droplet movement sliding on the surface of a vertical bare silicon wafer was regarded as the dynamic test for this imaging ellipsometric measurement system. This work was carried out at the incident angle of 70°, and 2 μl oil droplet

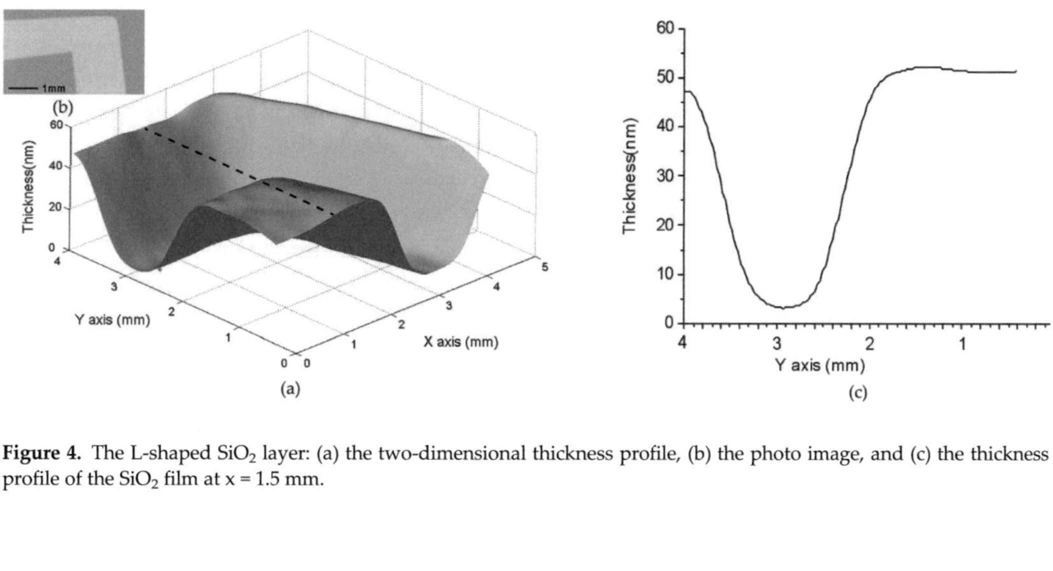


**Figure 4.** The L-shaped $SiO_2$ layer: (a) the two-dimensional thickness profile, (b) the photo image, and (c) the thickness profile of the $SiO_2$ film at x = 1.5 mm.

(Nikon, nd = 1.515) with high viscosity flowed slowly from the top of the vertically held silicon wafer. The total acquisition time of one set of ellipsometric parameters is about several tens of seconds, depending on the frame transfer speed of the camera. **Figure 5** demonstrates six sets of ellipsometric parameters during the oil dropping process.

### 3.3. Optimization for PEM Stokes imaging system

The Stokes parameters can be represented in the ellipsometric parameters by their definitions, as shown in **Figure 6**. If the input light is the +45° linear polarized light, the normalized output (reflected or transmitted) Stokes vector S = [$S_0$ $S_1$ $S_2$ $S_3$] can be expressed in the form of ellipsometric parameters as S = [1 cos(2Ψ) sin(2Ψ)cos(Δ) sin(2Ψ)sin(Δ)] [29]. Using the linear transformation model of polarimetry, we can write {b} = [A] {s}, where {b} is an N-element vector of the measured irradiances; [A] is an N × 4 matrix, the measurement matrix [30]; and {s} is the Stokes vector. Since {s} = $[A]^{-1}$ {b}, each element represents the response of the unit stimuli of the system. The noise in the measurement of the Stokes vector can be expressed in a vector form {n}; therefore, the error {ε} can be expressed as {ε} = $[A]^{-1}$ {n}. Since all components of the Stokes vector are weighted in noise production equally, the equally weighted variance (EWV) [29] figure of merit for N measurements can be expressed as follows:

$$EWV = \sum_{j=0}^{3}\sum_{k=0}^{N-1}\left([A]^{-1}\right)_{j,k}^{2} = Tr\left[[A]^{-1}\left([A]^{-1}\right)^{T}\right]. \tag{22}$$

This value demonstrates the measurement errors by summing all entries in the measurement matrix. The polarization state analysis portion of the PEM polarimetry consists of a photoelastic modulator and an analyser, whose azimuth angles were set at 0 and 45°, respectively. The kth row of the matrix $[A]^{-1}$ of the phase lock configuration can be expressed as $[1\ 0\ \cos(\Delta_P)\ \sin(\Delta_P)]^T$, where $\Delta_P = \delta_0 \sin\omega t$ is the phase retardation of the PEM. By taking the modulation amplitude to be half-wave (i.e., $\delta_o = \pi$), one can set the temporal phase at θ = ωt instead of moving the conventional rotating angles θ in space.

In this PEM polarimetry, the required minimum measurements for deducing Stokes vectors are 4. The EWV value for those temporal phases at 0, 30, 90, and 210° is 5, which is about one-quarter of the value for the classical rotating retarder and fixed polarizer system (RRFP) technique, as shown in **Figure 7**. Compared with the EWV value of the RRFP system under various configurations, one can observe that the noise is considerably reduced under the optimized phase retardation condition, but the angular positions have very limited effect on their EWV value. Since the EWV value can be used to quantitatively evaluate the noise immunity of a polarimetry, we can conclude that if one wants to achieve the same signal-to-noise ratio for 4 temporal phase measurements in the PEM polarimetry, one needs 8 measurements in the optimal orientations under optimal retardation and more than 16 uniformly spaced measurements over 360° in the RRFP configuration with quarter-wave retardation.

### 3.4. PEM imaging ellipsometric measurement with the dual-wavelength approach

In the previous section, two-dimensional ellipsometric parameters were determined at a specific wavelength. In the conventional PEM-based ellipsometric measurement, the modulation

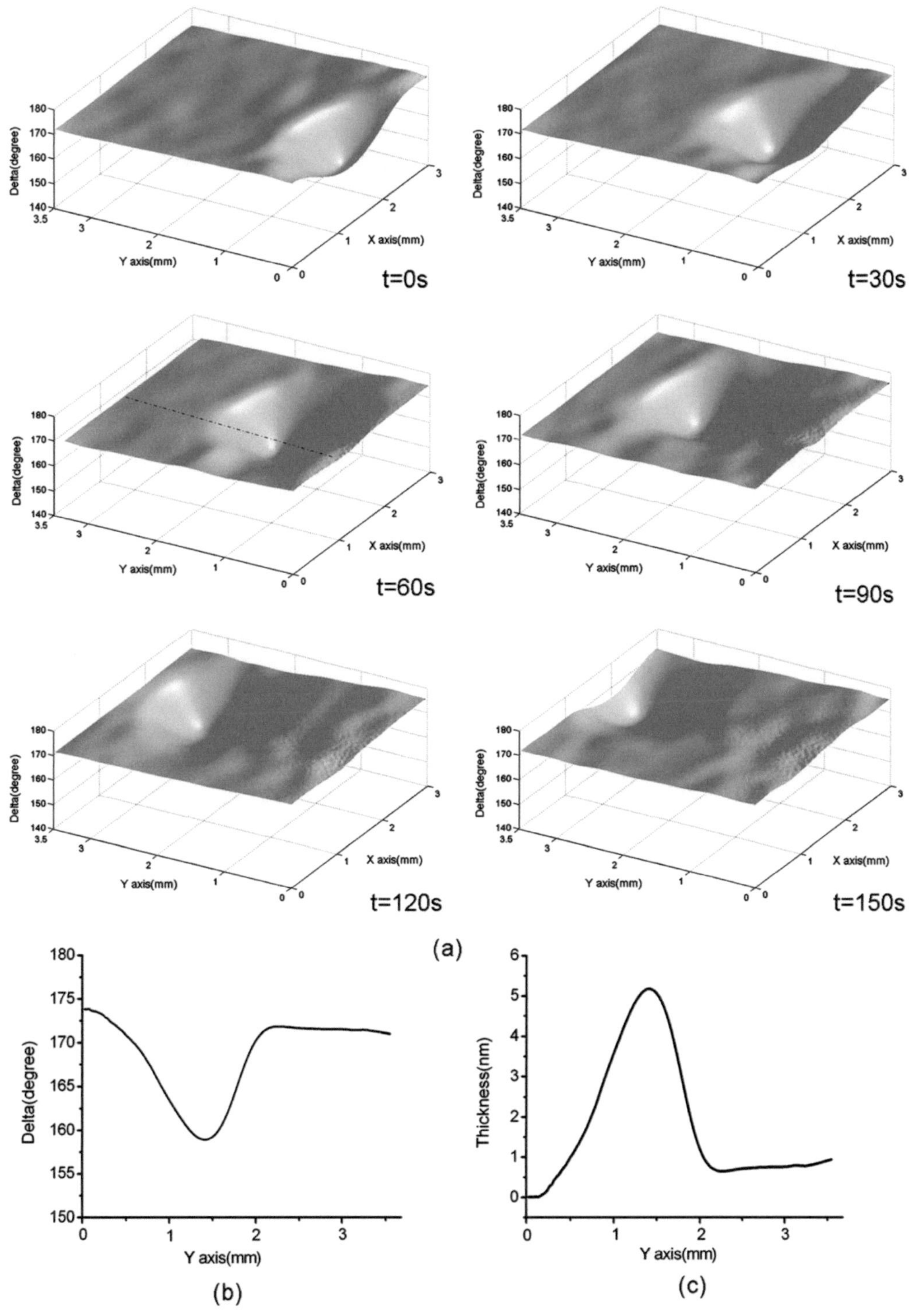


**Figure 5.** The movement behavior of an oil droplet: (a) Δ distribution for every 30 s, (b) the cross-sectional distribution of Δ through the center of the oil droplet at t = 60 s, and (c) the thickness profile across the center of the oil droplet at t = 60 s.

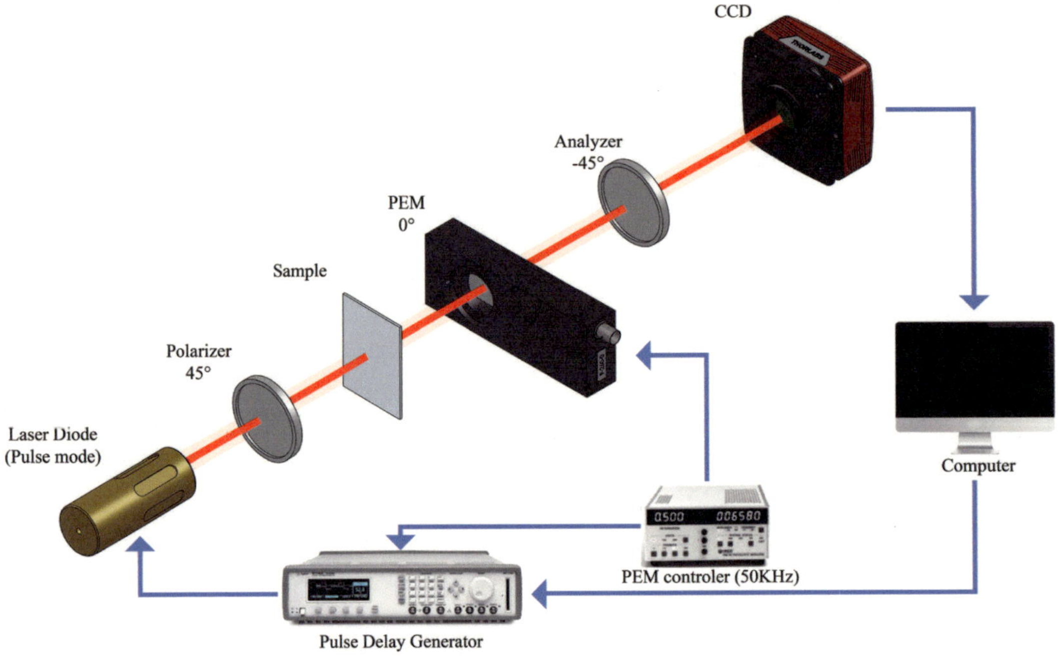


**Figure 6.** System configuration of the phase-lock PEM polarimetry for Stokes parameters measurement.

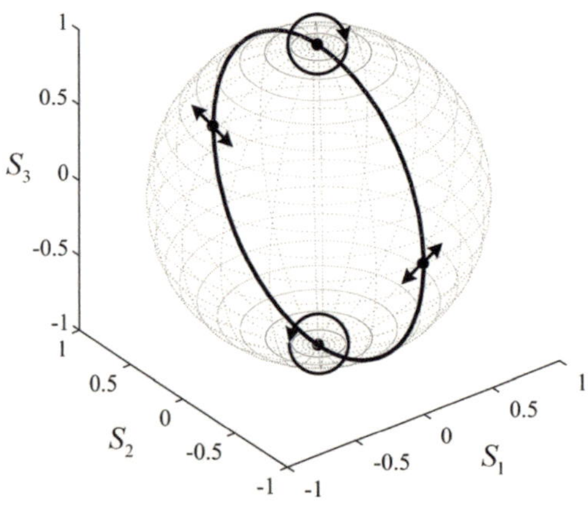


**Figure 7.** Trajectory of the phase-lock PEM polarimetry on the Poincar'e sphere: the four specific polarization states are indicated in the graph.

amplitude is controlled by applying an external voltage for a specific wavelength. However, a few seconds are required to reach resonance equilibrium and stabilize the modulator, while there is a change in the modulation amplitude for different wavelength; this procedure significantly reduced the measurement speed for the multi-wavelength measurement. Here, we developed an equivalent phase retardation technique that may help prevent the above disadvantage of PEM for multi-wavelength measurement [17]. In general, the half-wave modulation is set at a specific wavelength ($\lambda_1$). However, while the wavelength of the incident light shifts to the other shorter wavelength ($\lambda_2$), the modulation amplitude no longer equals 0.5 waves, but rather equals $0.5/\lambda_2$ waves. We changed the temporal phase angle $\omega t$ to maintain a constant dynamic retardation of the PEM, rather than the applied voltage, while the original wavelength $\lambda_1$ switches to the other $\lambda_2$. Their relation is as follows:

$$\Delta_p = \frac{0.5\lambda_1}{\lambda_2} \sin \omega t \tag{23}$$

According to the Eq. (22), **Figure 8** demonstrates that though the wavelength was changed from one to the other, the output polarizations were kept constant by setting different temporal phase angles within a modulation cycle. **Figure 9** shows the thickness profile of a two-step oxidized silicon wafer examined using red and blue light, respectively. **Table 1** lists the measured ellipsometric parameters and deduced film thickness from both wavelengths. To sum up, the film thicknesses measured were close to the theory whether by red or blue lights.

### 3.5. Full Mueller matrix imaging polarimetry based on the hybrid phase modulation

Mueller matrix imaging contains comprehensive information on the morphological and functional properties of the biological samples as well as the birefringence, dichroism, and depolarization of the specimens [31–33]. The conventional Mueller matrix imaging approaches were based on measurements involving sequential rotation of the polarizer, analyzer, and retarders,

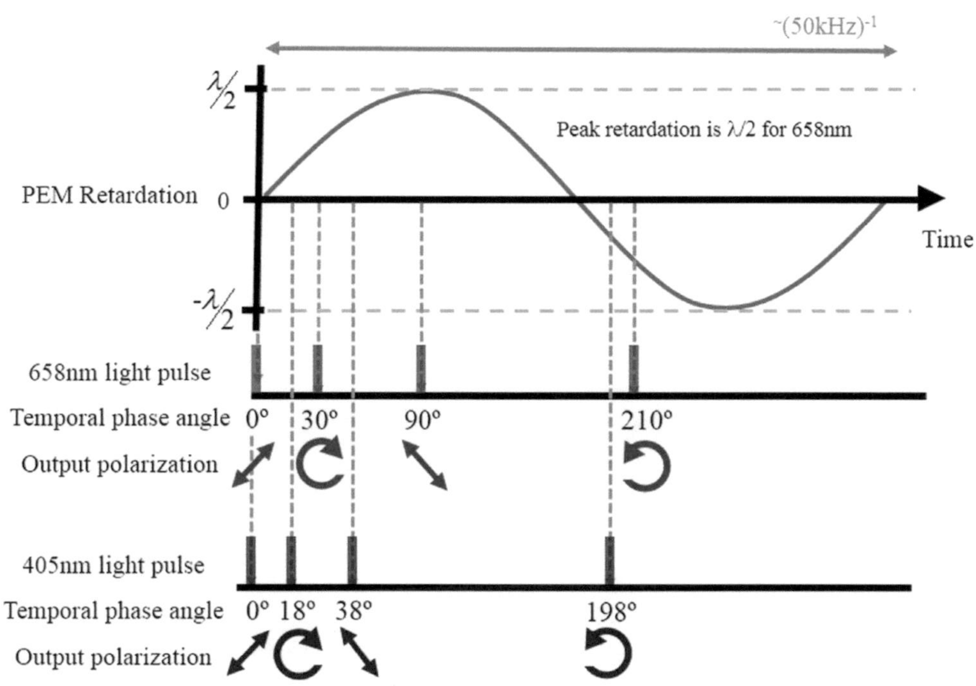


**Figure 8.** Polarization modulation at peak retardation of half-wave ($\lambda$ = 658 nm) and the output polarization at four temporal phase angles of dual-wavelength.

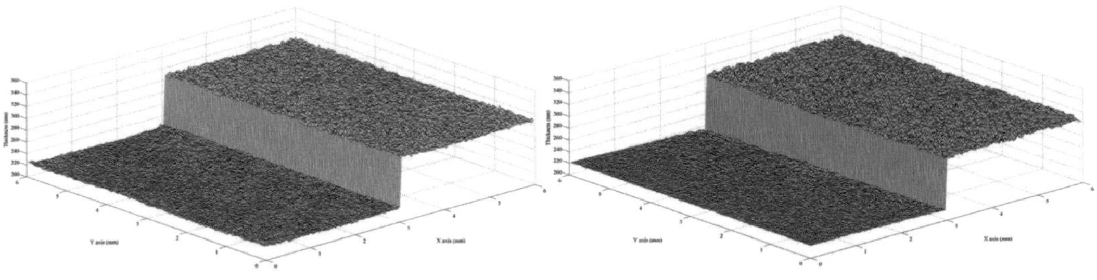

(a) (b)

**Figure 9.** Oxide thickness profile of the two-step reference wafer. (a) the deduced thickness profile by 658 nm red light source and (b) the deduced thickness profile by 405 nm blue light source.

| Wavelength (nm) | Index of refraction | Measured ellipsometric parameters | | Deduced film thickness (nm) |
|---|---|---|---|---|
| | | $\Delta(°)$ | $\Psi(°)$ | |
| 658 | Si:3.836-i0.016<br>$SiO_2$:1.456 | 166.03 ± 1.12 | 11.69 ± 0.08 | 5.20 ± 0.43 |
| 405 | Si:5.424-i0.330<br>$SiO_2$:1.469 | 159.72 ± 1.36 | 22.58 ± 0.38 | 5.57 ± 0.45 |

**Table 1.** Optical characteristics for the substrate and thin films.

but such approach was a time-consuming process that impeded the use of the system for in vivo imaging studies. In order to improve the speed of measurement, recent development in the Mueller matrix polarimeter was mainly focused on the process of using liquid crystal to control and analyze the state of the input or output polarizations. As the measurements with these approaches usually employed unmodulated light irradiance, the results were more sensitive to noise than methods using an intensity modulated light, particularly for highly scattering samples [34, 35]. Moreover, the tuning curve of the liquid crystal variable redarder (LCVR) was found to be sensitive to its alignment and temperature to result in systematic errors and also impact the overall performance of the instrument [36].

The deficiency of LCVRs in the Mueller matrix imaging system can be improved by replacing the LCVRs in the portion of the polarization state analyzer with a PEM, and the modified configuration is shown in **Figure 10** [18]. The polarization state generator of this system is composed of a linear polarizer and two LCVRs. The azimuth angle of the polarizer was set at $-45°$ and the slow axis of the two LCVRs was oriented to an angle of 90 and 45°, respectively. The retardations ($\delta_1$, $\delta_2$) of both LCVRs are dependent on their driving voltages to generate four polarizations. Thus, the Mueller matrix of the PSG in terms of the Mueller matrices of their components can be expressed as:

$$M_{PSG} = M_{LCVR_2}(\delta_2, 45°) \bullet M_{LCVR_1}(\delta_1, 90°) \bullet M_P(-45°) \tag{24}$$

The polarization state analyzer (PSA) was composed of a PEM and an analyzer. In order to obtain a complete set of PSA, we set the azimuth angle of the PEM at 0°, and the Mueller matrix representing the PSA module is obtained from:

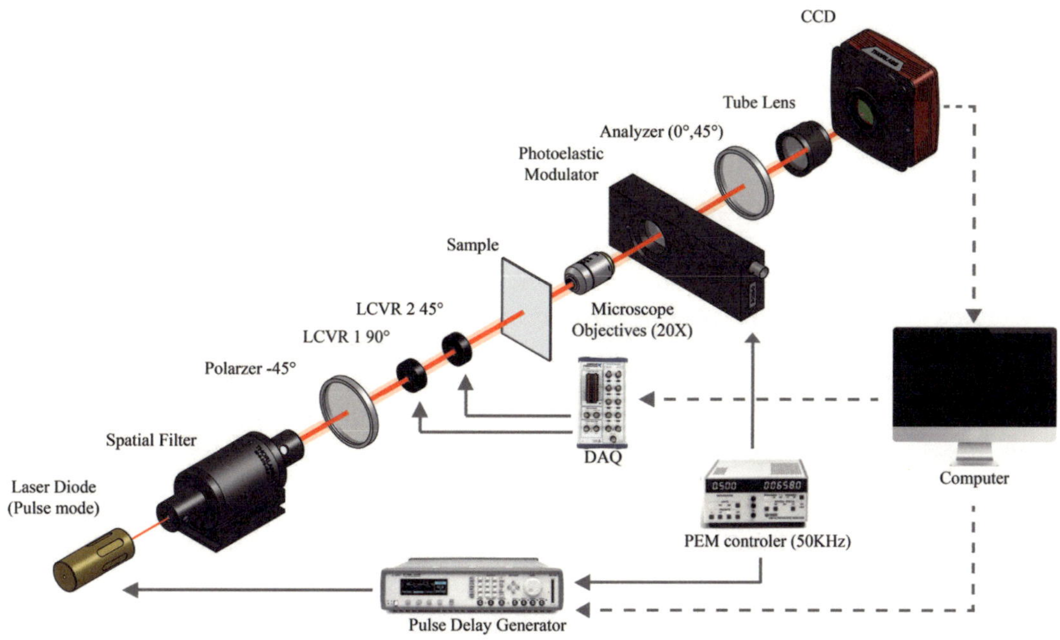


**Figure 10.** Optical setup for Mueller matrix imaging polarimetry with hybrid phase modulation technique.

$$M_{PSA} = M_A(A) \bullet M_{PEM}(\Delta_p, 0^\circ) \tag{25}$$

where $M_A$ and $M_{PEM}$ are the Mueller matrix of the analyzer and PEM, respectively; A is the azimuth angle of the analyzer; and $\Delta_p$ represents the phase retardation of the PEM, which was also modulated as $\delta_0 \sin\omega p$. Here, the amplitude of modulation $\delta_0$ is set at $\pi$, and the temporal phase angle refers to $\Delta_p$. Consequently, the total Mueller matrix of the system is given by:

$$M_T = M_{PSA}M_SM_{PSG} = M_a(A) \bullet M_{PEM}(\Delta_p, 0^\circ) \bullet M_S \bullet M_{LCVR_2}(\delta_2, 45^\circ) \bullet M_{LCVR_1}(\delta_2, 90^\circ) \bullet M_P(-45^\circ) \tag{26}$$

where Ms is represented as the Mueller matrix of the sample being tested. As the system is based on intensity modulation, only the first element of the Stokes parameters in Eq. (25) has to be considered. One can formulate the temporal intensity behavior as follows:

$$\begin{aligned} I(A, \theta_p, \delta_1, \delta_2) = \frac{I_0}{4} \langle & m_{00} - m_{02}\cos\delta_1 - \frac{1}{2} m_{01} \sin\delta_1 \sin\delta_2 \\ & + \cos 2A \left( m_{10} - \frac{1}{2} m_{11} \sin\delta_1 \sin\delta_2 - m_{12}\cos\delta_1 + m_{13}\sin\delta_1\cos\delta_2 \right) \\ & + \frac{1}{2} \sin 2A [\cos(\pi\sin\theta_p)(2m_{20} - m_{21}\sin\delta_1\sin\delta_2 - 2m_{22}\cos\delta_1) \\ & + \sin(\pi\sin\theta_p)(2m_{30}\sin 2A - m_{31}\sin\delta_1\sin\delta_2 - 2m_{32}\sin 2A)] \\ & + \sin\delta_1\cos\delta_2 \{m_{03} + \sin 2A[m_{23}\cos(\pi\sin\theta_p) + m_{33}\sin(\pi\sin\theta_p)]\} \rangle \end{aligned} \tag{27}$$

It is well-known principles that at least 16 individual polarization state measurements are required to determine the full Mueller matrix. The measurements are usually carried out by generating four specific polarization states from the PSG, and each output polarization can be determined by at least four intensity measurements. Accordingly, both the PSG and the PSA must be "complete" to obtain the full Mueller matrix with at least four basic states.

In detail, the four phase retardations ($\delta_1$, $\delta_2$) = (90°, 0°), (0°, 0°), (75.5°, 206.5°), and (75.5°, 153.5°) were sequentially set for both LCVRs, so that four specific polarization states were generated from the PSG. In order to characterize the Stokes vectors of the outgoing light from the sample, four conditions were also set up for the PSA by changing the azimuth of the analyzer A and the temporal phase angle $\Delta_p$ of the PEM, with the following conditions: (A, $\Delta_p$) = (0°, 0), (45°, 0°), (45°, 30°), and (45°, 90°). While capturing the images with different conditions of the PSG and the PSA, the modulated pulse is achieved by a DC bias current equal to the threshold value coupled with a programmable pulse generator to drive the laser diode. The generated and analyzed polarization states, in the order of the optimal optical settings, and the exact 16 intensity measurements were obtained as shown in **Table 2**, while the details for determining individual Mueller matrix elements are listed in **Table 3**.

Two results were shown by using this Mueller matrix imaging measurement system. The first results were the measured Mueller matrices of a quarter wave plate and the map of its phase retardation. We set the azimuth angle of the wave plate at 0°, which makes $m_{23}$ = 1, $m_{32}$ = -1, $m_{22}$ = 0, and $m_{33}$ = 0; the values of other elements were the same as those in air. Also, we rotated the quarter wave plate to set its azimuth angle at 30 and 60° and deduced its phase retardation by the Lu-Chipman algorithm, as shown in **Figure 11**. The average value of the phase retardation, which is close to the ideal condition at around 90°, and the azimuth angle under different rotation conditions are shown in **Figure 12**. Disregarding some static areas with small deviations due to speckles of dust in the imaging elements, the measured distributions almost matched the theoretical conditions.

The other result was the dynamic optical characteristics of a biopolymer specimen with heat-induced conformational change. The second test sample, shrimp shell, is composed of chitin, proteins, lipids, and pigments and with the characteristic of being semi-transparent. Accordingly, we investigated the conformational changes of shrimp shell induced by heat treatment,

| **PSG** ($\delta_1$, $\delta_2$) | **PSA (Analyzer, PEM)** | | | |
|---|---|---|---|---|
| | **(0°, 0°)** | **(45°, 0°)** | **(45°, 30°)** | **(45°, 90°)** |
| **(75.5°, 206.5°)** | $I_4$ | $I_5$ | $I_6$ | $I_7$ |
| **(75.5°, 153.5°)** | $I_3$ | $I_{10}$ | $I_9$ | $I_8$ |
| **(0°, 0°)** | $I_2$ | $I_{11}$ | $I_{12}$ | $I_{13}$ |
| **(90°, 0°)** | $I_1$ | $I_{16}$ | $I_{15}$ | $I_{14}$ |

**Table 2.** Measurement sequence of 16 intensities under the condition of retardation of LCVRs for the chosen set of analyzer and temporal phase angle of the PEM.

| Element | Intensity calculation |
|---|---|
| $m_{00}$ | $(2\cdot(2I_{10}-I_{11}-I_{13}+2I_5+2I_7+2I_8+2\sqrt{3}(I_{14}+I_{16})))/(2\sqrt{3}+3)$ |
| $m_{01}$ | $4\sqrt{3}\cdot(I_{10}-I_5-I_7+I_8)$ |
| $m_{02}$ | $(4\cdot(I_{10}-2I_{11}-2I_{13}+I_5+I_7+I_8-\sqrt{3}(I_{11}+I_{13}-I_{14}-I_{16})))/(2\sqrt{3}+3)$ |
| $m_{03}$ | $(2\cdot(2I_{10}-I_{11}-I_{13}-3I_{14}-3I_{16}+2I_5+2I_7+2I_8))/(2\sqrt{3}+3)$ |
| $m_{10}$ | $-(2\cdot(2I_{10}-I_{11}-I_{13}+2I_2-4I_3-4I_4+2I_5+2I_7+2I_8-4\sqrt{3}I_1+2\sqrt{3}(I_{14}+I_{16})))/(2\sqrt{3}+3)$ |
| $m_{11}$ | $4\sqrt{3}\cdot(I_{10}-2I_2+2I_4-I_5-I_7+I_8)$ |
| $m_{12}$ | $-(4\cdot(I_{10}-2I_{11}-2I_{13}+4I_2-2I_3-2I_4+I_5+I_7+I_8-2\sqrt{3}(I_1-I_2)-\sqrt{3}(I_{11}+I_{13}-I_{14}-I_{16})))/(2\sqrt{3}+3)$ |
| $m_{13}$ | $-(2\cdot(6I_1+2I_{10}-I_{11}-I_{13}-3I_{14}-3I_{16}+2I_2-4I_3+2I_5+2I_7+2I_8))/(2\sqrt{3}+3)$ |
| $m_{20}$ | $(2\cdot(2I_{10}-I_{11}+I_{13}+2I_5-2I_7-2I_8-2\sqrt{3}(I_{14}+I_{16})))/(2\sqrt{3}+3)$ |
| $m_{21}$ | $4\sqrt{3}\cdot(I_{10}-I_5+I_7-I_8)$ |
| $m_{22}$ | $(4\cdot(I_{10}-2I_{11}+2I_{13}+I_5-I_7-I_8-\sqrt{3}(I_{11}-I_{13}+I_{14}-I_{16})))/(2\sqrt{3}+3)$ |
| $m_{23}$ | $(2\cdot(2I_{10}-I_{11}+I_{13}+3I_{14}-3I_{16}+2I_5-2I_7-2I_8))/(2\sqrt{3}+3)$ |
| $m_{30}$ | $-(2\cdot(2I_{10}-I_{11}+2I_{12}-I_{13}+2I_5-4I_6+2I_7+2I_8-4I_9-4\sqrt{3}I_{15}+2\sqrt{3}(I_{14}+I_{16})))/(2\sqrt{3}+3)$ |
| $m_{31}$ | $4\sqrt{3}\cdot(I_{10}-I_5+2I_6-I_7+I_8-2I_9)$ |
| $m_{32}$ | $-(4\cdot(I_{10}-2I_{11}-2I_{13}+4I_{12}+I_5-2I_6+I_7+I_8-2I_9-2\sqrt{3}(I_{15}-I_{12})-\sqrt{3}(I_{11}+I_{13}-I_{14}-I_{16})))/(2\sqrt{3}+3)$ |
| $m_{33}$ | $-(2\cdot(2I_{10}-I_{11}+2I_{12}-I_{13}-3I_{14}+6I_{15}-3I_{16}+2I_5-4I_6+2I_7+2I_8-4I_9))/(2\sqrt{3}+3)$ |

**Table 3.** Set of 16 intensities for calculating the full Mueller matrix elements.

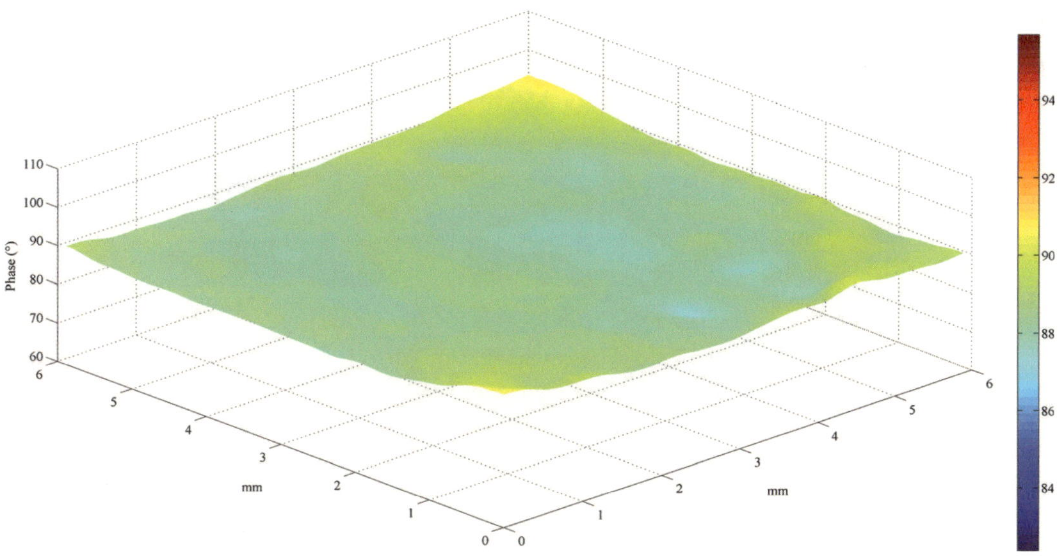


**Figure 11.** Phase distribution of the measured quarter wave plate while the azimuth angle was set at 0°.

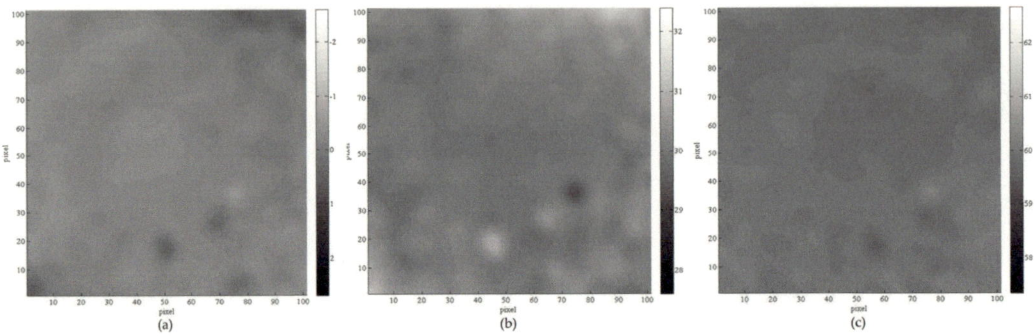


**Figure 12.** Azimuth angle distribution of the measured quarter wave plate while the azimuth angle was set at (a) 0°, (b) 30°, and (c) 60°.

because shrimp shell is a birefringent biopolymer material, whose absorption spectrum and transparency could be changed during the heat treatment. Therefore, the optical characteristics, such as diattenuation, depolarization, and phase retardation of the shrimp shell may be changed by heat treatment. During the examination, the shrimp shell was heated by a thermal electrical source to make it gradually turn to ruby red with the increase of the temperature. Meanwhile, a set of 16 images were recorded every 90 seconds to deduce the full elements of the Mueller matrix images. The retrieved polarization parameters, including retardance (R), depolarization coefficient (Δ), and diattenuation (d) deduced by the Lu-Chipman algorithm are shown in **Table 4**. Since the value of R was mainly related to the change of thickness or structure of the sample, the average value of R around 50° before and after the heat treatment of the shrimp shell represents thickness or structure that remained unchanged. Since the heat

|  | | Heating time | | | | | |
|---|---|---|---|---|---|---|---|
| | | 0 minute | 1.5 minute | 3 minute | 4.5 minute | 6 minute | 7.5 minute |
| Parameters | Δ | 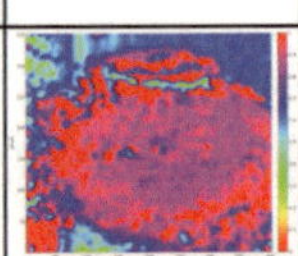 | 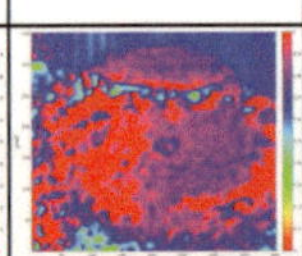 | 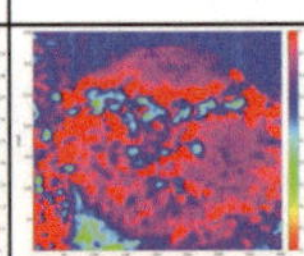 | 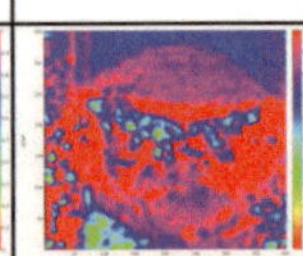 | 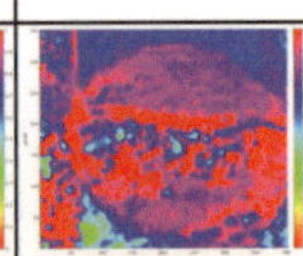 | 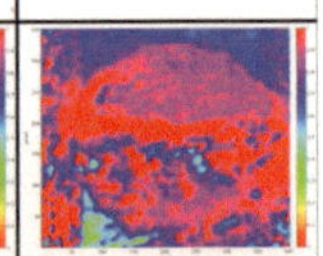 |
| | R(°) | 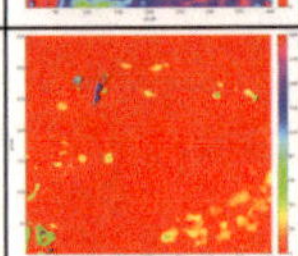 | 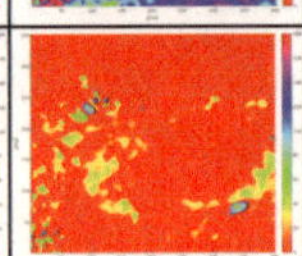 | 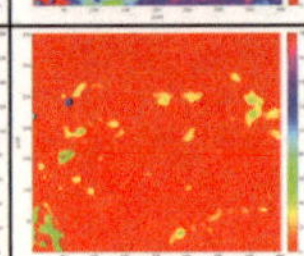 | 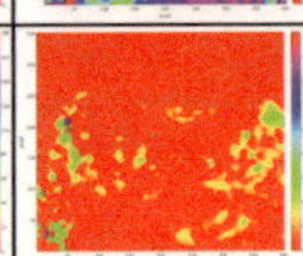 | 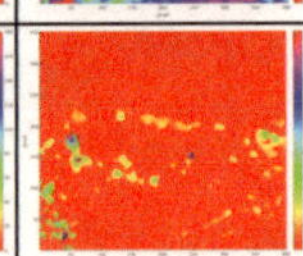 | 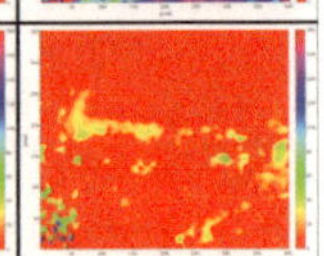 |
| | d | 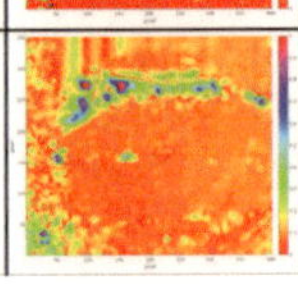 | 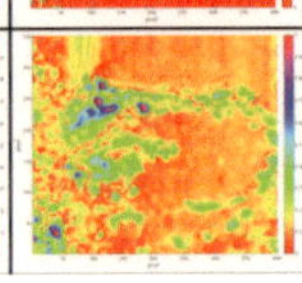 | 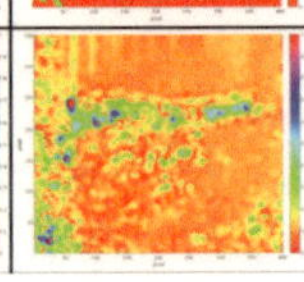 | 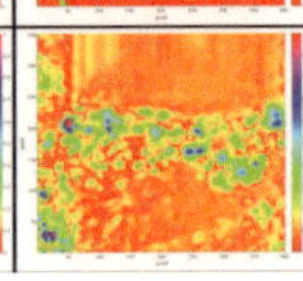 | 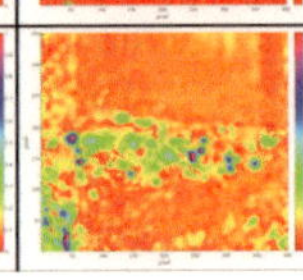 | 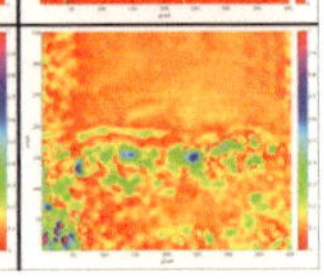 |

**Table 4.** Decomposed experimental Mueller matrix images of the shrimp shell by heating treatment.

source was located below the sample, the depolarization coefficient (Δ) appeared as the bottom-up change specification. One can observe that the overall trends for depolarization and diattenuation share common features due to heat treatment that induces the crustacean to relax its bonds with astaxanthin, which causes the shell to be transformed from almost transparent to red. In this process, the binding of both proteins would increase the scattering effect of the incident light. However, the behavior of the protein complex and transformation of the polarization properties associated with the Mueller matrix needs further exploration.

## 4. Conclusions

The known commercial imaging ellipsometry, usually employing the zone-averaging approach, is in the order of minutes to obtain two-dimensional optical characteristics of a thin film. In this chapter, we demonstrate the stroboscopic illumination technique in PEM-based ellipsometry just by using a limited number of measurements that can reduce the acquisition time in imaging ellipsometry and carry out the measurement within a few seconds. In addition to making use of this technique for imaging ellipsometry measurement, this approach was also extended to Stokes parameters and Mueller matrix imaging. Under the condition of multi-wavelength or spectroscopic measurement, polarimetry usually encounters the issue of retardation dispersion, which is the same situation that one encounters while using the PEM in an ellipsometry system. We developed the equivalent phase retardation technique, in which retardation dispersion settings of the PEM for different wavelengths were not required. That is to say, the stability and frame rate for the multi-wavelength measurement were almost the same as the single wavelength approach. The ellipsometric parameters of different wavelengths were capable of determining additional sample parameters, such as surface roughness, multiple film thicknesses, index dispersion, and the consistency of deduced results. Since polarization behavior of transmitted or reflected light was strongly related to their wavelength, multi-wavelength approach for Stokes parameters and Mueller matrix imaging can enhance the contrasts and increase the sensitivity of measurements for some features of the biological samples. If the polarimetry system can quickly obtain the desired images, we believe polarimetry imaging is well placed for in vivo tissue diagnosis in the forthcoming future [24].

## Author details

Chien-Yuan Han[1]*, Yu-Faye Chao[2] and Hsiu-Ming Tsai[3]

*Address all correspondence to: cyhan@gm.nuu.edu.tw

1 Department of Electro-Optical Engineering, National United University, Miaoli, Taiwan (ROC)

2 Department of Photonics, National Chiao Tung University, Hsinchu, Taiwan(ROC)

3 Department of Radiology, The University of Chicago, Chicago, USA

## References


[1] Tompkins HG. Industrial applications of spectroscopic ellipsometry. Thin Solid Films. 2004;**455–456**:772-778. DOI: 10.1016/j.tsf.2004.01.045

[2] Jin G, Meng YH, Liu L, Niu Y, Chen S, Cai Q, et al. Development of biosensor based on imaging ellipsometry and biomedical applications. Thin Solid Films. Amsterdam and New York, 2011;**519**:2750-2757. DOI: 10.1016/j.tsf.2010.12.175

[3] Fukuzawa K, Shimuta T, Nakada A, Zhang H, Mitsuya Y. Measurement of thickness of molecularly thin lubricant film using ellipsometric microscopy. IEEE Transactions on Magnetics. 2005;**41**:808-811. DOI: 10.1109/TMAG.2004.840349

[4] Azzam RMA, Bashara NM. Ellipsometry and Polarized Light. North Holland Publishing Co; 1978. p. 270-271. DOI: 10.1063/1.2994821

[5] Drévillon B. Phase modulated ellipsometry from the ultraviolet to the infrared: In situ application to the growth of semiconductors. Progress in Crystal Growth and Characterization of Materials. 1993;**27**:1-87. DOI: 10.1016/0960-8974(93)90021-U

[6] Asinovski L, Beaglehole D, Clarkson MT. Imaging ellipsometry: Quantitative analysis. Applications and Materials Science. 2008;**771**:764-771. DOI: 10.1002/pssa.200777855

[7] Qingqing L, Fukuzawa K, Kajihara Y, Zhang H, Itoh S. Development of ellipsometric microscope for high-resolution observation of nanometer-thick lubricant films. Tribology Online. 2011;**6**:251-256. DOI: 10.2474/trol.6.251

[8] Noort DV, Rumberg J, Jager EWH, Mandenius CF. Silicon based affinity biochips viewed with imaging ellipsometry. Measurement Science and Technology. 2000;**11**:801-808. DOI: 10.1088/0957-0233/11/6/325

[9] Chao YF, Lee KY. Index profile of radial gradient index lens measured by imaging ellipsometric technique. Japanese Journal of Applied Physics. 2005;**44**:1111-1114. DOI: 10.1143/JJAP.44.1111

[10] Han CY, Lee ZY, Chao YF. Determining thickness of films on a curved substrate by use of ellipsometric measurements. Applied Optics. 2009;**48**:3139-3143. DOI: 10.1364/AO.48.003139

[11] Graf RT, Eng F, Koenig JL, Ishida H. Polarization modulation fourier transform infrared ellipsometry of thin polymer films. Applied Spectroscopy. 1986;**40**:498-503. DOI: 10.1366/0003702864508818

[12] Vallon S, Compain E, Drévillon B. Improvements of fourier transform phase-modulated ellipsometry. Review of Scientific Instruments. 1995;**66**:3269-3272. DOI: 10.1063/1.1145492

[13] Khazimullin MV, Lebedev YA. Fourier transform approach in modulation technique of experimental measurements. Review of Scientific Instrumens. 2010;**81**:043110. DOI: 10.1063/1.3327844

[14] Han CY, Chao YF. Photoelastic modulated imaging ellipsometry by stroboscopic illumination technique. Review of Scientific Instruments. 2006;**77**:023107. DOI: 10.1063/1.2173027

[15] Tsai HM, Chen CW, Tsai TH, Chao YF. Deassociate the initial temporal phase deviation provided by photoelastic modulator for stroboscopic illumination polarization modulated ellipsometry. Review of Scientific Instruments. 2011;**82**:035117. DOI: 10.1063/1.3568745

[16] Tsai HM, Chao YF. Optimization of a four-temporal phase lock for photoelastic-modulated polarimetry. Optics Letter. 2009;**34**:2279-2281. DOI: 10.1364/OL.34.002279

[17] Kuo CW, Han CY, Jhou JY, Peng ZY. Using a fast dual-wavelength imaging ellipsometric system to measure the flow thickness profile of an oil thin film. Applied Surface Science. http://www.sciencedirect.com/science/article/pii/S0169433217301010. DOI: 10.1016/j.apsusc.2017.01.100

[18] Han CY, Du CY, Jhou JY. Rapid full Mueller matrix imaging polarimetry based on the hybrid phase modulation technique. Optics Communication. 2017;**382**:501-508. DOI: 10.1016/j.optcom.2016.08.026

[19] Wang B, List J. Basic optical properties of the photoelastic modulator: Part I. Useful aperture and acceptance angle. Proceedins of SPIE. 2005;**5888**:58881I. DOI: 10.1117/12.617904

[20] Wang B, Hinds E, Krivoy E. Basic optical properties of the photoelastic modulator part II: Residual birefringence in the optical element. Proceedings of SPIE. 2009;**7461**:746110. DOI:10.1117/12.826392

[21] Zeng A, Li F, Zhu L, Huang H. Simultaneous measurement of retardance and fast axis angle of a quarter-wave plate using one photoelastic modulator. Applied Optics. 2011;**50**:4347-4352. DOI: 10.1364/AO.50.004347

[22] Gribble A, Layden D, Vitkin IA. Experimental validation of optimum input polarization states for Mueller matrix determination with a dual photoelastic modulator polarimeter. Optics Letters. 2013;**38**:5272-5275. DOI: 10.1364/OL.38.005272

[23] Alali S, Yang T, Vitkin IA. Rapid time-gated polarimetric Stokes imaging using photoelastic modulators. Optics Letters. 2013;**38**:2997-3000. DOI: 10.1364/OL.38.002997

[24] Alali S, Vitkin A. Polarized light imaging in biomedicine: Emerging Mueller matrix methodologies for bulk tissue assessment. Journal of Biomedical Optics. 2015;**20**:61104. DOI: 10.1117/1.JBO.20.6.061104

[25] Alali S, Gribble A, Vitkin IA. Rapid wide-field Mueller matrix polarimetry imaging based on four photoelastic modulators with no moving parts. Opt Lett.2016;**41**:1038–1041. DOI: 10.1364/OL.41.001038

[26] Wang M, Chao Y, Wang M, Chao Y, Leou K, Chao YF, et al. A direct determination technique for Azimuth alignment in photoelastic modulation ellipsometry. Japanese Journal of Applied Physics. 1999;**38**:6919-6922. DOI: 10.1143/JJAP.38.6919

[27] Wang MW, Chao YF. Azimuth alignment in photoelastic modulation ellipsometry at a fixed incident angle. Japanese Journal of Applied Physics. 2002;**41**:3981-3986. DOI: 10.1143/JJAP.41.3981

[28] Wang MW, Chao YF, Leou KC, Tsai FH, Lin TL, Chen SS, et al. Calibrations of phase modulation amplitude of photoelastic modulator. Japanese Journal of Applied Physics. 2004;**43**:827-832. DOI: 10.1143/JJAP.43.827

[29] Collett E. Polarized Light: Fundamentals and Applications. New York: Marcel Dekker; 1993. p. 581

[30] Sabatke DS, Descour MR, Dereniak EL, Sweatt WC, Kemme SA, Phipps GS. Optimization of retardance for a complete Stokes polarimeter. Optics Letters. 2000;**25**:802-804. DOI: 10.1364/OL.25.000802

[31] Chen C, An I, Ferreira GM, Podraza NJ, Zapien JA, Collins RW. Multichannel Mueller matrix ellipsometer based on the dual rotating compensator principle. Thin Solid Films. 2004;**455–456**:14-23. DOI: 10.1016/j.tsf.2003.11.191

[32] Dubreuil M, Rivet S, LeJeune B, Cariou J. Snapshot Mueller matrix polarimeter by wavelength polarization coding. Optics Express. 2007;**15**:13660-13668. DOI: 10.1364/OE.15.013660

[33] Gao M, Yang P, Mckee D, Kattawar GW. Mueller matrix holographic method for small particle characterization: Theory and numerical studies. Applied Optics. 2013;**52**:5289-5296. DOI: 10.1364/AO.52.005289

[34] Laude Boulesteix B, De Martino A, Drévillon B, Schwartz L. Mueller polarimetric imaging system with liquid crystals. Applied Optics. 2004;**43**:2824-2832. DOI: 10.1364/AO.43.002824

[35] De Martino A, Kim YK, Garcia Caurel E, Laude B, Drévillon B. Optimized Mueller polarimeter with liquid crystals. Optics Letters. 2003;**28**:616-618. DOI: 10.1364/OL.28.000616

[36] Babilotte P, Nunes Henrique Silva V, Sathaye K, Dubreuil M, Rivet S, Dupont L, et al. Twisted ferroelectric liquid crystals dynamic behaviour modification under electric field: A Mueller matrix polarimetry approach using birefringence. Japanese Journal of Applied Physics. 2014;**115**:34906. DOI: 10.1063/1.4862162

Chapter 7

# Spectroscopic Ellipsometry - Application on the Classification of Diamond-Like Carbon Films

XiaoLong Zhou and Hidetoshi Saitoh

Additional information is available at the end of the chapter

http://dx.doi.org/10.5772/intechopen.71727

**Abstract**

Diamond-like carbon (DLC) films have been spreading from their theoretical basis to worldwide industrial applications because of their unique properties. Since their properties depend strongly on the conditions of synthesis, the effective classification of DLC films becomes quite necessary. From the ternary phase diagram to the Japan New Diamond Forum standard, the classification attempts are also accompanied by the continuous development of their applications. Generally, the hydrogen content and $sp^3/(sp^2 + sp^3)$ ratio are the primary parameters for their classification. However, researchers are afraid that currently $sp^3/(sp^2 + sp^3)$ ratio estimated included not only network $sp^3$ but also $sp^3$ hybrid carbons in the hydrogen-terminated cluster. Simultaneously, the above classification methods need to use the large equipment, such as the synchronous radiation source. Therefore, to realize more straightforward to classify DLC films efficiently, the optical constants (refractive index ($n$) and extinction coefficient ($k$)) have been proposed in 2013 to be effective method to classify the DLC films, for which a lot of considerable discussion in the past ISO/TC-107 meetings has been made. The purpose of this chapter is to introduce the latest developments of optical constants on the classification of DLC films and explore their relationship with the current standard.

**Keywords:** spectroscopic ellipsometry, diamond-like carbon film, classification

## 1. Introduction

Diamond-like carbon (DLC) film is one of the attractive carbon materials due to its outstanding properties which have wide applications in mechanical, electrical, optical, and chemical fields [1]. In general, DLC films have complex structures composed of the "diamond-like" $sp^3$ hybrid carbons C($sp^3$), $sp^2$ hybrid carbons C($sp^2$), and hydrogen terminations [2, 3]. Many structural models have been proposed to describe the role of each component in DLC films.

One of the most famous models is the "two-phase structure model" which is proposed by Robertson [4]. This model considers that the DLC structure can be described as the C($sp^2$) clusters embedded in the matrix of C($sp^3$). The hydrogen terminations also play such an important role in all types of DLC films. Previous studies have emphasized that the bonding states of hydrogen atoms in the DLC films have been C($sp^3$)-H in many cases [5]. When the hydrogen content exceeds a certain amount, the C($sp^3$)-H bonds become close to the polyethylene (PE) structure where hydrogen atoms make the bonds with the $sp^3$ hybridized carbon atoms. Thus, the PE structure should also be taken into account as another phase of the DLC structural model. Then, a "three-phase model" might be appropriate to describe the DLC films which have certain hydrogen contents. With the continuous expansion of research, the effective classification of DLC films based on structural analysis becomes critical, because their properties depend strongly on the conditions and methods of synthesis [6]. In order to classify the DLC films systematically, various experimental methods have been used for structural analysis of DLC films. Raman spectroscopy, X-ray photoelectron spectroscopy, electron energy loss spectroscopy, near-edge X-ray absorption fine-structure (NEXAFS), solid-state nuclear magnetic resonance techniques, and Rutherford backscattering and elastic recoil detection analysis (RBS/ERDA) have been used to obtain the $sp^3/(sp^3 + sp^2)$ ratios or hydrogen contents of DLC films [3, 7, 8]. Based on these studies, the researchers in Germany and Japan have proposed a classification standard of DLC films in 2005 [9] and 2012 [10], successively. In this classification, DLC films are categorized into the four typical groups, so-called DLC films, type I : tetrahedral amorphous carbon (ta-C), type II: hydrogenated tetrahedral amorphous carbon (ta-C:H), type III: amorphous carbon (a-C), and type IV: hydrogenated amorphous carbon (a-C:H). And also other two special types, such as type V: graphite-like carbon (GLC) films and type VI: polymer-like carbon (PLC) films.

Nowadays, not only the structural analysis but also the physical, chemical, and mechanical properties have been used to distinguish the types of DLC films. Especially, the optical constants in terms of the refractive indices ($n$) and extinction coefficients ($k$) (at $\lambda = 550$ nm) measured by spectroscopic ellipsometry (SE) have been proved to be a simple and effective method to classify the DLC films. It was proposed in 2013 [11], for which a lot of considerable discussion in the past ISO/TC-107 meetings in 2016 (BSI, UK) and 2017 (Tokyo, Japan) has been made [12]. Since this method needs to use only an SE without the large-scale equipment (the hydrogen content was measured by the RBS/ERDA using an electrostatic accelerator [13], and the $sp^2/(sp^2 + sp^3)$ ratio was obtained by NEXAFS based on synchrotron radiation [2, 7, 14]), it is more conducive to simplify the classification of DLC films. However, it is necessary to discuss the consistency between the structural analyses made with SE and NEXAFS. In addition, Bruggeman effective medium approximation (BEMA) has been widely used to analyze the structure of DLC films [15, 16]. The BEMA theory is made up from the Lorentz-Lorenz equation based on a Lorentz local field theory and assumes the spherical dielectrics distributed in the mixed phases. In the BEMA equation, the dielectric function of DLC films ($\varepsilon_a$) can be approximated in terms of those of the selected specific standard materials. In the previous studies, graphite, highly oriented pyrolytic graphite (HOPG = PG in this chapter), glassy carbon (GC), diamond, PE, and void have been used as the standard materials. The $\varepsilon_a$ can be obtained from the BEMA model simulations. Then, it is possible to analyze the structure of DLC films by fitting the spectrum measured by SE. However, the results of the previous analyses have been strongly dependent on the kind of the standard materials and their dielectric functions, and in addition, the types of DLC films have been limited to only a few species. Therefore, it is necessary to find the well-defined standard materials and the dielectric function of each standard material through the analysis of various types of DLC films.

In this chapter, we discuss the application of SE in the current classification of DLC films through two parts of experiments. In part one, DLC films were assumed to be represented by the superposition of the standard materials, for which the BEMA theory was applied to reproduce the experimental result of SE. From these analyses, the different types of DLC films can be represented by varying the superposition coefficients of the standard materials. We have selected five standard materials to build ten kinds of optical models after considering the feasibility of each model. In order to verify the reliability of the results, we used the X-ray reflectivity (XRR) method to measure the film densities as well as the NEXAFS method to evaluate the $sp^3/(sp^3 + sp^2)$ ratios and then compared the densities and ratios with that calculated from BEMA theory for individual films [14]. Finally, the discussion and comparison of the above methods are made on the structural analysis of various types of DLC films. In part two, we refurbished the classification of amorphous carbon films based on the optical constants ($n$ and $k$). In the selected photon energy range, we defined the maximum of $n$ ($E_{n\text{-max}}$) and $k$ at a value more than $10^{-4}$ ($E_k$) to explore the relationship between the different classification schemes for amorphous carbon films deposited by different techniques [17].

## 2. Experimental details

In the part one experiment, 13 DLC films were prepared by different deposition methods which included samples provided by the other research groups upon the request by the authors. All the samples were deposited on a (100)-oriented p-type single silicon wafer substrate. Samples #01–03 were the provided samples which were deposited by a filtered cathodic vacuum arc (FCVA) method, and the detail conditions were unclear (to obtain ta-C films). Samples #04–06 were made by a FCVA method, where the graphite target (The Nilaco Co., Ltd., purity of 99.9%) was used as the carbon precursor at a negative bias voltage in the range of 1.0–2.0 kV, a deposition time of 10 min with a working pressure of 0.5 Pa, and an arc voltage of 800 V (to obtain a-C or ta-C:H films). Samples #07 and #08 were prepared by a radio-frequency (RF) magnetron sputtering method at a negative bias voltage of 0.3 kV, a deposition time of 5 and 3 min with a working pressure of 10 Pa, and a RF power of 150 W (to obtain ta-C:H or a-C:H films). Samples #09 and #10 were deposited by an electron-cyclotron-resonance chemical vapor deposition (ECRCVD) method at a negative bias voltage of 0.3 and 0.5 kV, a deposition time of 10 min with a working pressure of 0.5 Pa, and an RF power of

100 W (to obtain a-C:H or PLC films). Samples #11–13 were prepared by a plasma-enhanced (PE)-CVD method at the same 10-min deposition time and applied negative bias voltage in the range of 0.0–0.5 kV (to obtain a-C:H or PLC films). In the part two experiment, various types of DLC films were also deposited on a (100)-oriented p-type single silicon wafer substrate. Samples A–C were prepared by the FCVA deposition which is provided by the other research group upon the request of the authors. Samples D and E were deposited by RF magnetron sputtering at a negative bias voltage of 0.3 kV, a deposition time of 3 and 5 min with a working pressure of 20 Pa, and an RF power of 150 W. Samples F–K were synthesized by RF-PE-CVD methods at the same 10-min deposition time and applied negative bias voltage in the range of 0.0–0.5 kV.

All the DLC samples and the selected standard materials of PG and GC plates were measured with an ellipsometer (HORIBA, Jobin-Yvon, UVISEL NIR 23301010I). The incident angle of the source radiation was set to 70°; each measurement was carried out in the spectral range between 0.6 and 4.8 eV with a step of 0.05 eV at 293 K. The values of Tauc-band gap ($E_g$), $n$, and $k$ are obtained from the fitting of the experimental data [18]. XRR was used to determine the true density and thickness of standard material GC and all the DLC films in the part one. In this approach, the X-ray was incident on the surface of the above materials at a small angle (0.18° < $2\theta$ < 2.0°), and the total reflection occurs at a critical angle. The XRR measurements (M03XHFMXP3, Mac Science) were taken by using a Cu $K\alpha$ source with the wavelength of 1.54 Å operated at an acceleration voltage of 40 kV and its currents of 15 mA under the condition of the scan range of 0.18–2.00° and the step size of 0.004°. The hydrogen content was evaluated by RBS/ERDA with 2.5 MeV $He^+$ irradiation using an electrostatic accelerator (Nisshin-High Voltage, NT-1700HS) located at the Extreme Energy Density Research Institute, Nagaoka University of Technology. The $sp^3/(sp^3 + sp^2)$ ratios of DLC films were obtained by NEXAFS method which was performed at the BL3.2Ub of the Synchrotron Light Research Institute (SLRI) (Public Organization), Nakhon Ratchasima, Thailand. The NEXAFS spectra were measured in the partial electron yield (PEY) mode, and the light polarization was parallel to the surface at any incident light angle. The total energy resolution was approximately 0.5 eV. The carbon $K$-edge NEXAFS spectra were measured in the energy range of 275–320 eV at an energy step of 0.1 eV. Absolute photon energy was obtained by adjusting the $\pi^*$ (C = C) peak position of graphite. Uncertainty in the calibrated energy was estimated to be ±0.2 eV.

## 3. SE-BEMA analysis

### 3.1. Selection of standard materials

As described above, the main constituents in the structure of DLC films are C($sp^2$), C($sp^3$), and hydrogen terminations. Therefore, we selected the standard materials having these three basic structures individually as PG, GC, diamond, PE, and void. PG plate (NT-MDT, ZYA quality, GRAS/1.0) with the mosaic spread of 0.3–0.5° was used as the approximate crystalline

http://dx.doi.org/10.5772/intechopen.71727

C(*sp*[2]) standard material. GC (Tokai Fine Carbon Co., Ltd., Grade GC-20SS) was used as the noncrystalline C(*sp*[2]) standard material which was mirror-polished, and the surface roughness is 20 nm. We selected crystal diamond as the crystalline C(*sp*[3]) standard material. Although several researchers [13, 14] used the $sp^3$-amorphous carbon to approximate the structure of C($sp^3$), it was difficult to define the dielectric function and density of $sp^3$-amorphous carbon. Nonetheless, we used the dispersion equation of Sellmeier's formula (extinction coefficient, $k = 0$), [19]. For the structure of hydrogen terminations, previous studies pointed out that hydrogen atoms generally form $-CH_n$ ($n = 1$, 2, and 3) groups with C($sp^3$) which can be simulated by PE. We selected PE as the noncrystalline C($sp^3$) standard material. The dispersion equation of classical oscillator model [20] used for the dielectric function of PE. In addition, the density and dielectric constant of DLC films were generally lower than those of crystalline diamond and graphite because of the amorphization and hydrogenation. This is due to the free volume formed during the film deposition process. For these reasons, we selected the void as one of the standard materials. The meaning of void has a wide variety, such as porosity and empty space. Guo et al. [21] have given the definition of the void as the existence of the free space resulting from incomplete interlocking of nuclei. There was no specific reference material so that we consider the permittivity of void approximately equal to the atmosphere as $\varepsilon_r = 1.0$ and $\varepsilon_i = 0$.

### 3.2. Optical model establishment

SE is an optical technique for investigating the dielectric properties of thin films, from which the film properties such as thickness, roughness, optical constants, electrical conductivity, compositions, and other material properties can be obtained. SE is an indirect method, that is, the measured amplitude component ($\Psi$) and phase difference ($\Delta$) cannot, in general, be converted directly into the optical constants of a sample. Usually, a model analysis must be performed by using the models of Lorentz, Drude, classical dispersion, new amorphous, Tauc-Lorentz (TL) dispersion [18, 22], etc., which are used to fit and estimate the complex refractive index (dielectric function) of different materials. In these models, the Tauc-Lorentz dispersion formula has been the most commonly used to calculate the dielectric function of amorphous materials. The Levenberg–Marquardt algorithm has been used to check the fitting analysis of the ($\Psi$, $\Delta$) spectra. The fitting error ($\chi^2$) of each measurement was obtained from the difference between the experimental and simulated ($\Psi$, $\Delta$) values [23]. The final result was obtained so as the $\chi^2$ was considered to be the smallest in each fitting analysis. To determine the volume fraction and film thickness, the BEMA optical models were established to perform the simulation analyses based on the SE ($\Psi$, $\Delta$) spectra. Considering that the model should be commonly applied to standard materials and all types of DLC films, we used the dispersion formula that combines Tauc-Lorentz (TL) and Drude models [23] (TL + Drude) to simulate the measured ($\Psi$, $\Delta$) spectra. **Figure 1** shows all the optical models of standard material (**Figure 1(a)** for PG and (b) for GC) and DLC films (**Figure 1(c)–(l)**).

The simulation analyses of standard materials (PG and GC) were carried out before that of DLC films. For the PG, as shown in **Figure 1(a)**, the substrate was modeled by TL + Drude

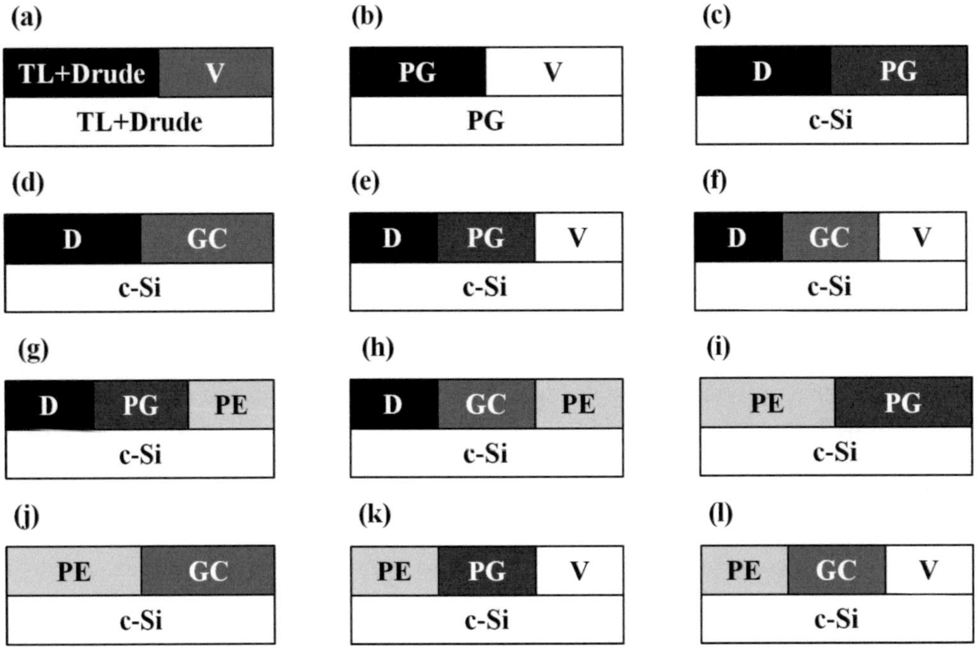


**Figure 1.** Optical models applied to BEMA equation (TL + Drude models, PG = HOPG, GC: glassy carbon, D: diamond, PE: polyethylene, and V: void). The models (a) and (b) are used for the simulation of PG and GC standard materials, the models (c)–(h) are used for the simulation of all types of DLC films, and the models (i)–(l) are used for the high hydrogen DLC films.

and the layer on the substrate was (TL + Drude) + void (50%). For the simulation of GC, which is considered to consist a certain amount of void and three-dimensional structural $sp^2$ hybrid carbon, the PG + void (50%) optical model with the BEMA screening factor $q = 0.333$ was used to fit the observed ($\Psi$, $\Delta$) spectrum of the GC plate (roughness at 20 nm order). The dielectric function of PG and GC from these simulation analyses is shown in **Figure 2**. **Figure 2(a)** shows the $\Psi$ and $\Delta$ spectra obtained for the measurement of PG plate and their simulated spectra by the dispersion equation of optical model (a). The observed spectra were fully reproduced by the simulated spectra, from which $\chi^2$ of this simulation was 2.641, indicating that this optical model is appropriate to evaluate the dielectric function of PG. **Figure 2(b)** shows the comparison of the dielectric functions of PG obtained from the above analysis and Ref. [24]. As shown in **Figure 2(b)**, the present $\varepsilon_r$ value shows any undulation at 0.9 eV which is observed in Ref. [24]. From the above results, we consider that the present dielectric function of PG can be used to build up the other optical models. **Figure 2(c)** shows the simulation of the GC based on the present dielectric function of PG. The $\chi^2$ of the simulation was 4.845, which is larger than that of PG. Nonetheless, the observed and simulation data showed a good agreement. The comparison of the calculated dielectric functions and those of the GC reference [25] is shown in **Figure 2(d)**, from which it is indicated that the present dielectric functions of GC are closer to the reference functions than those of PG described above. The results suggest that our

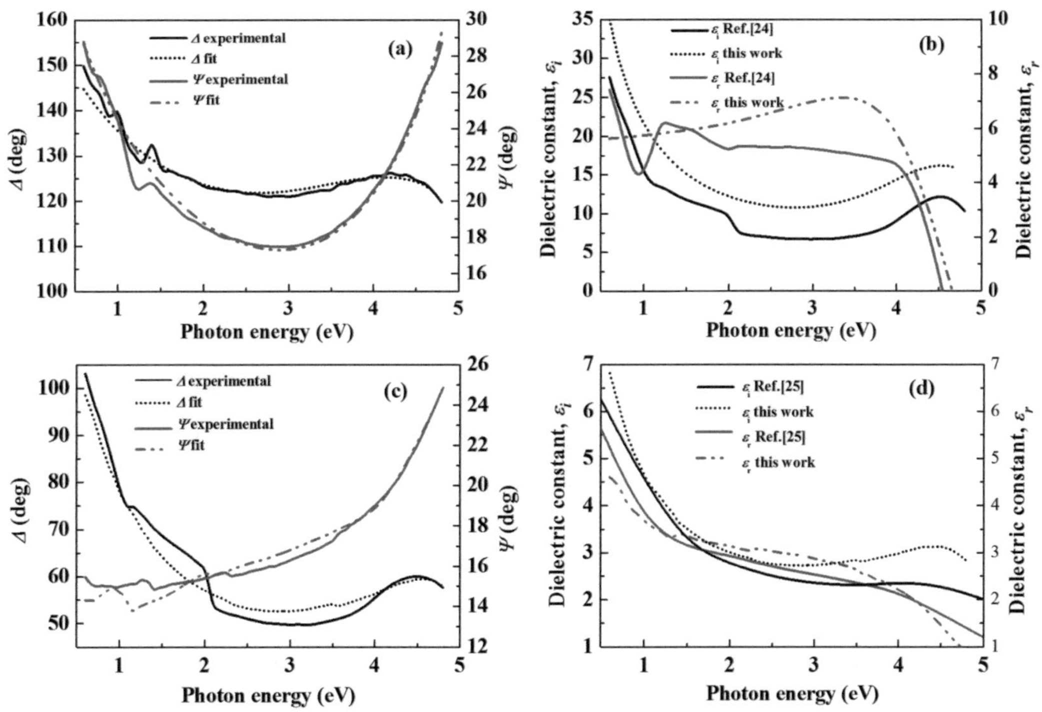


**Figure 2.** Simulation analysis of (Ψ, Δ) spectra and the resultant dielectric factions of PG and GC measured at the angle of incidence of 70. In (a) and (c), the solid line is the experimental Ψ and Δ spectra, and the dotted line is simulated Ψ and Δ spectra, respectively. In (b) and (d), the solid line is $\varepsilon_r$ and $\varepsilon_i$ functions, respectively, quoted from Ref. [24] for PG and Ref. [25] for GC, and the dotted line is those obtained from the present chapter.

analysis of SE using the BEMA theory and the selection of the standard materials of PG and GC are feasible in this chapter.

In the simulation of DLC films, we have assumed ten optical models as shown in **Figure 1(c)–(l)**. The crystalline silicon (c-Si) was used as the substrate in each optical model. The first layer of each optical model was assumed to consist of a mixed phase which was composed of several standard materials. The diamond or PE was assumed to be the standard materials of C($sp^3$) component as well as PG or GC for the standard materials of the C($sp^2$) component. For the free volume or hydrogenated termination, void or PE was assumed as standard materials in these optical models. Optical model (c)–(f) which have no PE phase contain with low hydrogen content should be suitable for types I, III, and V DLC films, especially, optical model (c) and (d) more suitable for ta-C or a-C films. Optical model (g)–(l) have the PE phase contain with hydrogen content in some degree should be suitable for types II, IV, and VI DLC films, especially optical models (k) and (l) more suitable for PLC films. In the first stage of these simulation analyses, all the DLC films were simulated by using the models of (c)–(h) where the diamond is one of the basic standard materials. However, the standard material of diamond cannot necessarily reflect the C($sp^3$) structure for all kinds of the present DLC films, which can, instead, be represented well

| Standard material | PG | GC | Diamond | PE | void |
|---|---|---|---|---|---|
| Density (g/cm³) | 2.26 | 1.69 | 3.52 | 0.95 | 0 |

**Table 1.** Densities of standard materials used for the analysis of the BEMA density.

by the PLC films when they are fabricated by PECVD or ECRCVD methods. In the second stage, therefore, the models (i)–(l) using PE as one of the basic standard materials are adapted to analyze the high-hydrogen content and low-density DLC films. In order to verify the correctness of the volume fractions of the individual components obtained by BEMA method and to validate the optical models used for the individual DLC samples, we calculated the BEMA density ($\rho_{BEMA}$) as:

$$\rho_{BEMA} = \sum_{i=1}^{3} \rho_i \times f_i. \tag{1}$$

In Eq. (1), $\rho_i$ is the density of the standard material and $f_i$ is the volume fraction. The $\rho_i$ values are summarized in **Table 1**. Since the density of GC has been scarcely reported [26], we obtained the density of GC from the XRR measurement of a GC plate which will be described later. While the density of PE has a wide range of 0.91–0.96 g/cm$^3$ [27], the density of 0.95 g/cm$^3$ was used in this chapter which is closer to the high-density PE and DLC films.

## 4. Results and discussion

### 4.1. Results

In the part one experiment, the commercial software (Rigaku, GXRR) was used to simulate the logarithmic data of the reflection intensity, from which the density and thickness of GC and DLC films can be obtained. **Figure 3** shows the XRR profiles of the standard material of GC on a semilogarithmic scale. The solid curve represents the experimental profile, and the dotted curve the simulation analysis. In the high angle side (> 0.5°), there is a large discrepancy between the simulation and experimental curves. Nonetheless, the true density obtained from XRR method is generally determined by the critical angle, whose simulated value 0.358° is in good agreement with the experimental value 0.360°. The true density can be determined by the critical angle, and the thickness analyses by the fringe pattern which are obtained with GXRR software based on the Parrat's method [28]. Therefore, the discrepancy mentioned above in the large-angle region has no effect on the determination of the GC density. From the above measured critical angle of GC plate, the density is estimated to be 1.67 g/cm$^3$. Another measurement point yielded the density to be 1.71 g/cm$^3$. Then, the final density of standard GC was the arithmetic average of the two densities as 1.69 g/cm$^3$ in this chapter.

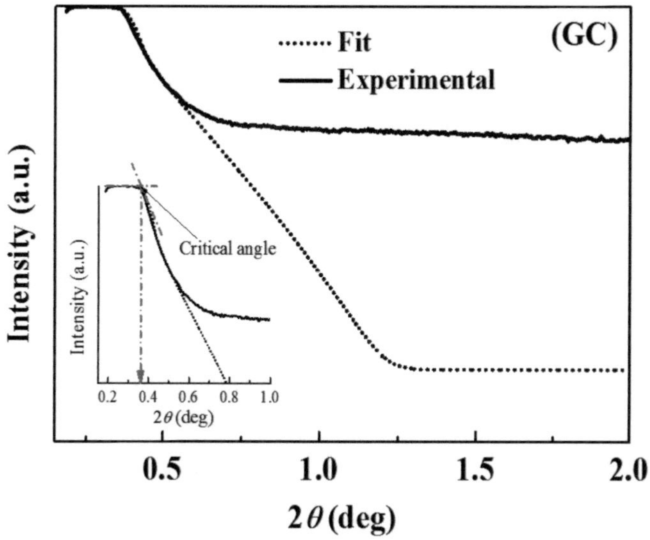


**Figure 3.** XRR profiles of GC (solid line: experimental profile, dotted line: fitted profile).

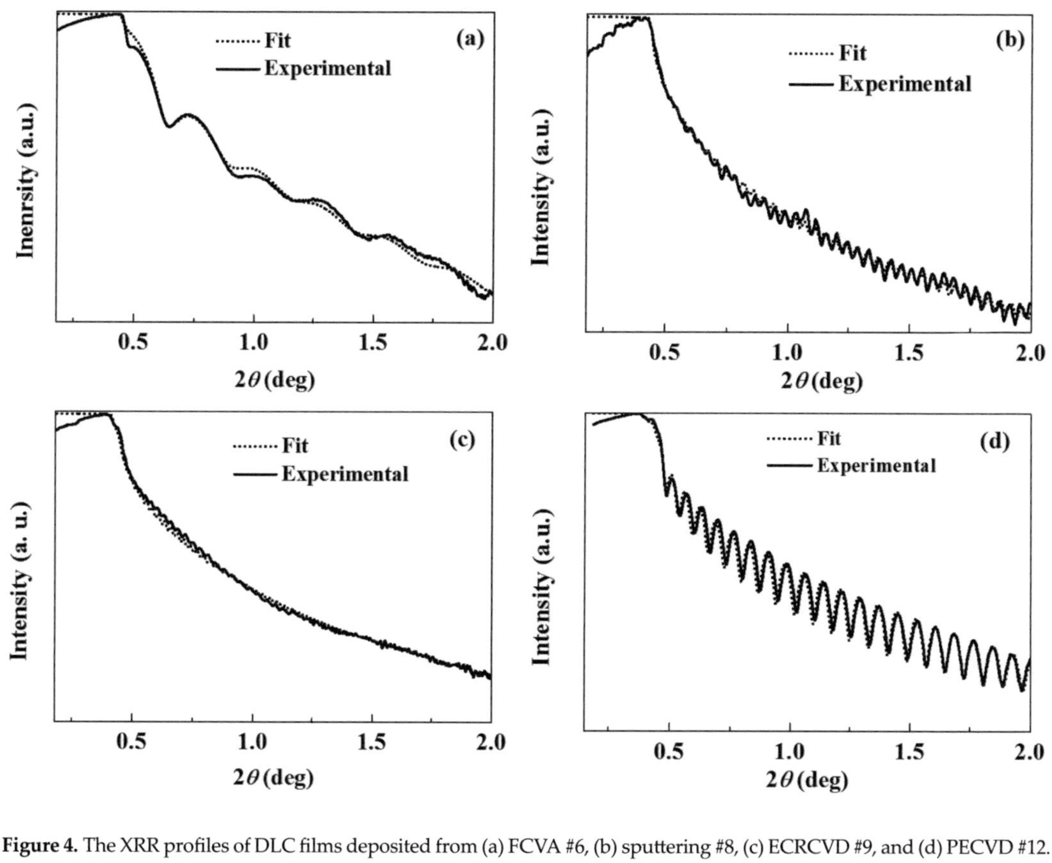


**Figure 4.** The XRR profiles of DLC films deposited from (a) FCVA #6, (b) sputtering #8, (c) ECRCVD #9, and (d) PECVD #12.

The examples of the XRR profiles and simulation results for the samples of #06, #08, #09, and #12 made by FCVA, sputtering, ECRCVD, and PECVD methods, respectively, are shown in **Figure 5**. The critical angles of these samples were 0.43, 0.40, 0.38, and 0.35°, from which the XRR densities ($\rho_{XRR}$) were 2.62, 2.21, 1.98, and 1.50 g/cm$^3$, respectively. Since the interferogram (the fringe in each XRR profile) appears periodically to change with the film thickness, the XRR thickness ($d_{XRR}$) of each film obtained from the fitting was 21.4, 188.8, 256.0, and 220.0 nm.

For all the other samples, $\rho_{XRR}$ and $d_{XRR}$ have also been obtained by the same way, and all the results are summarized in **Table 2**. Due to the variations of the deposition methods and the deposition time, $d_{XRR}$ of deposited DLC films in this chapter is distributed between 21.4 and 312.0 nm. In addition, the film density was dependent on the film formation method. From these densities, the present DLC films can be classified as follows. The ϱXRR values of DLC films made by FCVA-I method are in the range of 3.02–3.25 g/cm$^3$, those of obtained from FCVA-II method are in the range of 2.01–2.62 g/cm$^3$, those of made by sputtering method are in the range of 2.17–2.21 g/cm$^3$, and those obtained from the ECRCVD and PECVD methods are in the range of 1.23–2.00 g/cm$^3$.

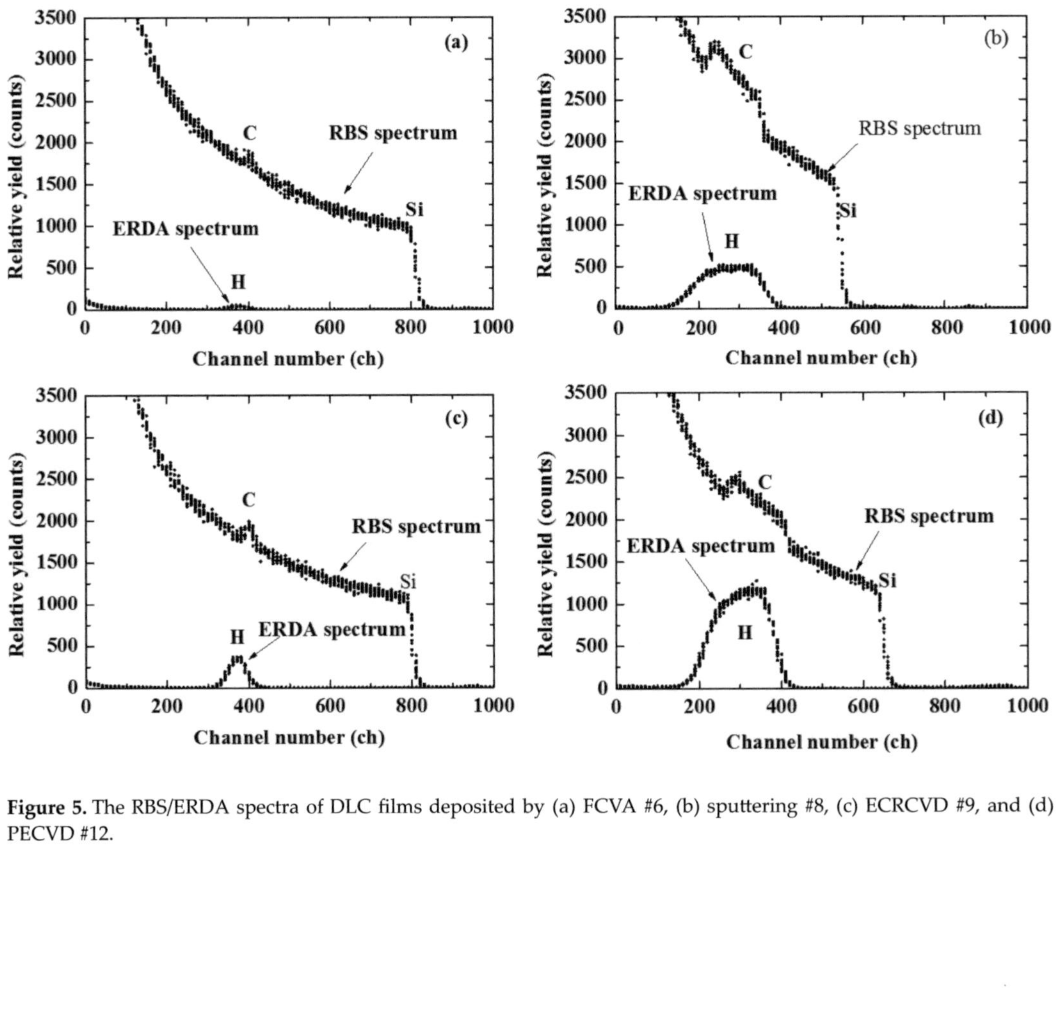


**Figure 5.** The RBS/ERDA spectra of DLC films deposited by (a) FCVA #6, (b) sputtering #8, (c) ECRCVD #9, and (d) PECVD #12.

| Sample | Method | $d_{XRR}$ (nm) | $\rho_{XRR}$ (g/cm³) | $\rho_{BEMA}$ (g/cm³) | H content (at.%) |
|---|---|---|---|---|---|
| #01 | FCVAI | 181.0 | 3.10 | 3.23 | 0.3 |
| #02 | FCVAI | 66.3 | 3.02 | 3.14 | 0.5 |
| #03 | FCVAI | 210.8 | 3.25 | 3.17 | 1.0 |
| #04 | FCVAII | 29.8 | 2.53 | 2.30 | 4.5 |
| #05 | FCVAII | 26.9 | 2.01 | 2.01 | 4.5 |
| #06 | FCVAII | 21.4 | 2.62 | 2.39 | 5.0 |
| #07 | Sputtering | 312.0 | 2.17 | 2.15 | 19.0 |
| #08 | Sputtering | 188.8 | 2.21 | 2.26 | 19.0 |
| #09 | ECRCVD | 256.0 | 2.00 | 2.05 | 26.0 |
| #10 | ECRCVD | 250.0 | 1.26 | 1.19 | 33.0 |
| #11 | PECVD | 104.8 | 1.74 | 1.73 | 31.0 |
| #12 | PECVD | 220.0 | 1.50 | 1.45 | 36.0 |
| #13 | PECVD | 50.7 | 1.23 | 1.23 | 42.0 |

**Table 2.** Summary of $d_{XRR}$, $\rho_{XRR}$, $\rho_{BEMA}$, and hydrogen content of the DLC films.

**Figure 5** shows the typical RBS and ERDA spectra of DLC films deposited by (a) FCVA #06, (b) sputtering #08, (c) ECRCVD #09, and (d) PECVD #12 methods. For the RBS spectra, the peaks that $He^+$ ions backscattered according to carbon atoms in DLC films are observed around 200–450 ch. For the ERDA spectra, the peaks of hydrogen atoms recoiled from the sample by the irradiation of $He^+$ ions emerged around 180–420 ch. The C and Si peaks on RBS spectra of the DLC films and the substrates are profiled using an RBS fitting calculation package. The H peaks on ERDA spectra of the DLC films are profiled using an ERDA fitting calculation package to compare the peak intensities of C and H. The estimating error of the present fitting process is around 5%. The hydrogen contents of DLC films of #06, #08, #09, and #12 were 5.0, 19.0, 26.0, and 36.0 at.%, respectively. **Table 2** lists the other results obtained from the above peak fittings. The hydrogen content of DLC film of type made by FCVA method was in the range of 0.3–1.0 at.% and that of type in the range of 4.5–5.0 at.%. The hydrogen content of DLC films obtained from sputtering method was 19.0 at.%. The hydrogen contents obtained from the ECRCVD and PECVD methods were in the range of 26.0–42.0 at.%.

**Figure 6** shows the examples of the carbon *K*-edge NEXAFS spectra of DLC films synthesized by (a) FCVA #06, (b) sputtering #08, (c) ECRCVD #09, and (d) PECVD #12 methods. These spectra are decomposed into the two-edge structures. A pre-edge resonance near 285 eV is assigned to the transition from the C 1 *s* orbital to the unoccupied $\pi^*$ orbitals principally originating from $sp^2$ or *sp*. sites if present. The other broadband between energy edges from 288 to 320 eV is related to the C 1 $s \rightarrow \sigma^*$ transitions at the *sp*., $sp^2$, and $sp^3$ sites

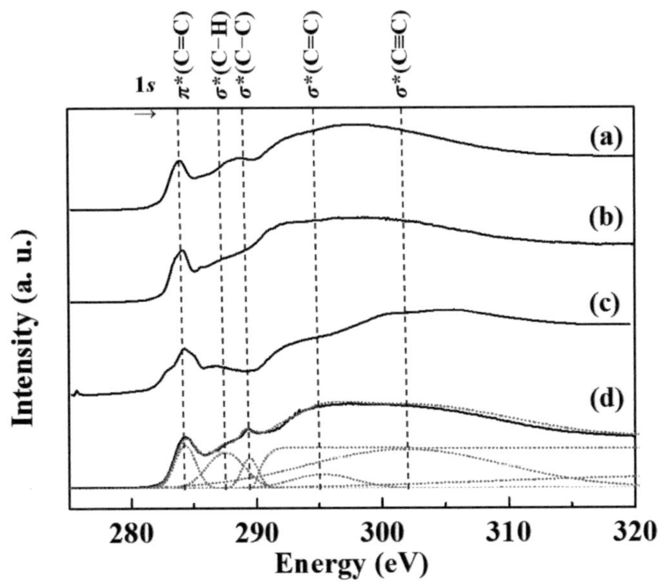


**Figure 6.** The carbon *K*-edge of NEXAFS spectra of DLC films deposited by (a) FCVA #6, (b) sputtering #8, (c) ECRCVD #9, and (d) PECVD #12.

in the DLC films. The $sp^2$ content of the DLC films can be extracted by normalizing the region of the resonance corresponding to the C 1 $s \rightarrow \pi^*$ at 284.6 eV with the whole spectral area. The absolute $sp^2$ content was evaluated by the purely $sp^2$ reference measured on the PG carbon *K*-edge as follows from which the $sp^3/(sp^3 + sp^2)$ ratio can be evaluated. The estimating error of the present fitting process is also around 5%. The $sp^3/(sp^3 + sp^2)$ ratios of DLC films were evaluated as listed in **Table 3** from the curve fitting of the NEXAFS spectra shown in **Figure 6(d)**.

The commercial software (HORIBA DelaPsi2) was used to simulate the SE data of standard materials and DLC films. **Figure 7** shows the typical fitting results of Ψ and Δ spectra of SE method with the application of the BEMA equations with different optical models. In the first stage, the optical models (c)–(h) using diamond as one of the basic standard materials were applied to the DLC films. The simulation results of sample #06 which has low-hydrogen content are shown in **Figure 7(a)** and **(b)**. The solid lines represent the experimental data, and the dashed lines the fitted data in each optical model. Except for $\Psi$ of the model (c), the other simulation results were in good agreement with the experimental data. The $\chi^2$ values of these simulations were 4.53, 0.93, 0.58, 0.89, 0.61, and 1.13, for the models of (c)–(h). The BEMA densities ($\rho_{BEMA}$) calculated by Eq. (1) using the models of (c)–(h) were 3.32, 2.39, 2.08, 2.32, 1.65, and 2.27 g/cm$^3$, respectively. The calculated $\rho_{BEMA}$ of sample #06 using model (d) was close to the value of $\rho_{XRR}$. **Figure 7(c)** and **(d)** show the simulation of the $\Psi$, $\Delta$ spectra obtained from sample #11 using the same models (c)–(h), where the simulated spectra showed the large deviation from the experiments. The $\chi^2$ of these fittings were 15.23, 4.15, 1.06, 0.84, 1.94, and 0.92 for the models of (c)–(h), respectively. The calculated $\rho_{BEMA}$

| Sample/group | | Film type | H (at.%) | $sp^2/(sp^2+sp^3)$ (%) | $n$ | $k$ | $E_{n\text{-max}}$ (eV) | $E_k$ (eV) | $E_g$ (eV) | $d$ (nm) | $\rho$ (g/cm³) |
|---|---|---|---|---|---|---|---|---|---|---|---|
| | | | | | $\lambda = 550$ nm | | | | | | |
| A | I | *ta*-C | 0.3 | 47.9 | 2.65 | 0.13 | 3.45 | 0.70 | 0.68 | 207 | 3.25 |
| B | I | *ta*-C | 0.5 | 45.6 | 2.66 | 0.22 | 2.55 | 0.65 | 0.62 | 66 | 3.14 |
| C | I | *ta*-C | 1.0 | 44.3 | 2.75 | 0.30 | 2.00 | 0.75 | 0.71 | 181 | 3.23 |
| D | II | *a*-C:H | 19 | 56.6 | 2.34 | 0.31 | 1.60 | 0.65 | 0.61 | 312 | 2.17 |
| E | II | *a*-C:H | 19 | 59.1 | 2.42 | 0.29 | 1.90 | 0.80 | 0.76 | 186 | 2.21 |
| F | III | *a*-C:H (PLC) | 31 | 58.5 | 2.17 | 0.17 | 2.40 | 1.05 | 0.71 | 491 | 1.73 |
| G | III | *a*-C:H (PLC) | 32 | 67.1 | 2.15 | 0.19 | 2.20 | 1.00 | 0.89 | 501 | 1.72 |
| H | III | *a*-C:H (PLC) | 36 | 62.0 | 2.04 | 0.07 | 2.65 | 1.30 | 1.05 | 476 | 1.49 |
| I | III | *a*-C:H (PLC) | 37 | 64.9 | 1.97 | 0.05 | 2.30 | 1.15 | 0.92 | 457 | 1.43 |
| J | III | *a*-C:H (PLC) | 41 | 57.5 | 1.86 | 0.05 | 3.35 | 1.95 | 1.36 | 350 | 1.35 |
| K | IV | PLC | 45 | 63.0 | 1.65 | 0.00 | 4.80 | 2.60 | 2.55 | 254 | 1.21 |

**Table 3.** Hydrogen contents, $sp^2/(sp^2 + sp^3)$ ratios, $n$, $k$, $E_{n\text{-max}}$, $E_k$, $E_g$, thickness ($d$), and true density ($\rho$) of DLC films, estimated from the analysis of RBS/ERDA, NEXAFS, SE spectra, and XRR profiles.

were 3.16, 2.42, 1.65, 1.73, 1.53, and 1.39 g/cm³. Sample #11 obtained the optimal value of $\chi^2$ in the case of model (f). Since samples #10–13 had high-hydrogen content, in the second stage, optical models (i)–(l) are PE adopted as the basic standard material and used to simulate the samples. The representative results of sample #13 are shown in **Figure 7(e)** and **(f)**. All the simulation data showed a good agreement with the experimental data (**Figure 7(e)** need to consider flipping up the raw data at part of 2.5–4.0 eV range to compare). The $\chi^2$ of these fitting results using models (i)–(k) were 0.37, 0.36, and 0.33. The $\rho_{\text{BEMA}}$ in these cases were 1.33, 1.20, and 1.23 g/cm³. In addition, the volume fraction in the model (l) showed a negative value, being neglected in this chapter. $\rho_{\text{BEMA}}$ of sample #13 has close agreement with $\rho_{\text{XRR}}$ in the case of model (k). All the other $\rho_{\text{BEMA}}$ and $\rho_{\text{XRR}}$ of each sample are listed in **Table 2**.

In part two experiment, the RBS/ERDA, XRR, and NEXAFS analysis are also performed on all samples. All the results are summarized in **Table 3**. The film thicknesses are in the range of 66–501 nm. Samples A–C are hydrogen free (under 1.0 at.%) DLC films with high true density (exceed 3.10 g/cm³). Samples D and E have the same hydrogen content of 19 at.% and almost the same true density of 2.20 g/cm³. As the substrate bias decreases from 0.5 to 0 kV, the hydrogen contents of samples E and F increase from 31 to 45 at.% and their densities decrease from 1.73 to 1.21 g/cm³. The estimated errors of these present fitting process are around 5%

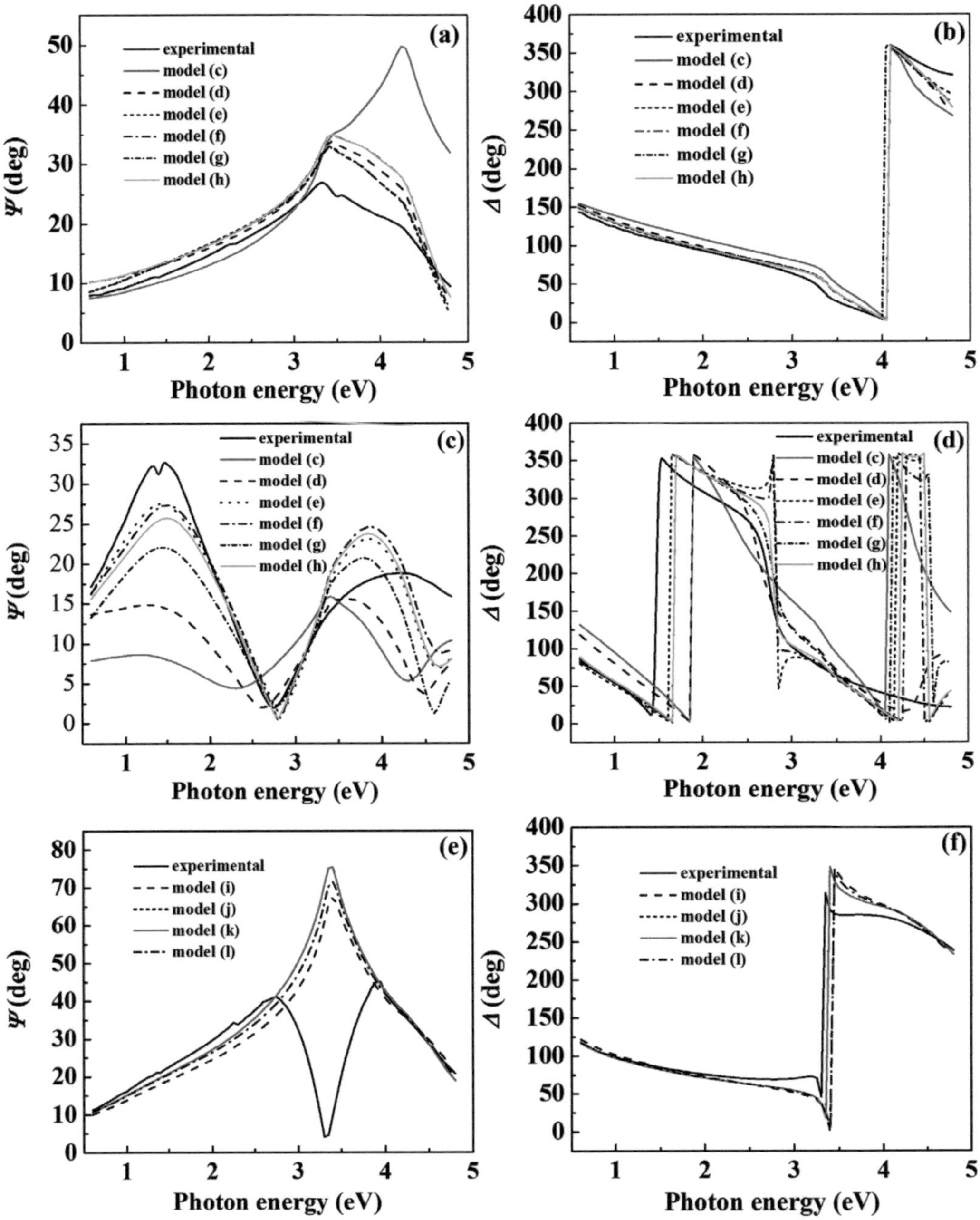


**Figure 7.** Representative simulation results of Ψ and Δ spectra of SE method applied BEMA equation with first six optical models (models (c)–(h) for FCVA #6 (a, b), PECVD #11 (c, d) methods). The other four optical models (models (i)–(l) for PECVD #13 (e, f) methods).

(e.g., sample F is 31 ± 2 at.%), ± 0.05 g/cm$^3$, and ±5 nm, respectively. According to our previous classification scheme, the present films can be well classified into three types: t*a*-C, *a*-C:H, and PLC films [14], although the boundary of hydrogen content between *a*-C: H and PLC films is still controversial.

**Figure 8(a)** shows the results of spectra of $n$ in the photon energy range of 0.6–4.8 eV which is evaluated by using a three-layer optical model based on the Tauc-Lorenz and effective medium approximation model theory [11]. These films can be categorized into four groups based on the $n$ at the energy of 2.25 eV ($\lambda$ = 550 nm). Group includes samples A–C, where $n$ is in the range of 2.65–2.75 and classified as t*a*-C films. Group includes samples D and E, where $n$ is in the range of 2.34–2.42 and classified as *a*-C:H films. The difference between the minimum value of $n$ in group and the maximum value of $n$ in group is 0.17 and between its minimum value in group and maximum value in sample F is 0.16. In addition, the difference of $n$ between samples J and K is 0.16, which is in the same degree with the above-mentioned differences. The difference between any other two samples is less than 0.1. Therefore, the sample K is categorized into groups, classified as PLC films in **Table 3**; the other samples are categorized into groups, classified as *a*-C:H or PLC films. Because the present grouping shows a good agreement with the classification of DLC films based on the hydrogen content, $sp^2/(sp^2 + sp^3)$ ratio, and true density, we validate that the gap between the four groups has a practical significance on the classification of DLC films.

Except for sample K with a maximum value of $n$ at ≥4.8 eV [29], all the DLC films show the $E_{n\text{-max}}$ in the range of 0.6–4.8 eV as shown in **Figure 8(a)**. In this section, $E_{n\text{-max}}$ is defined

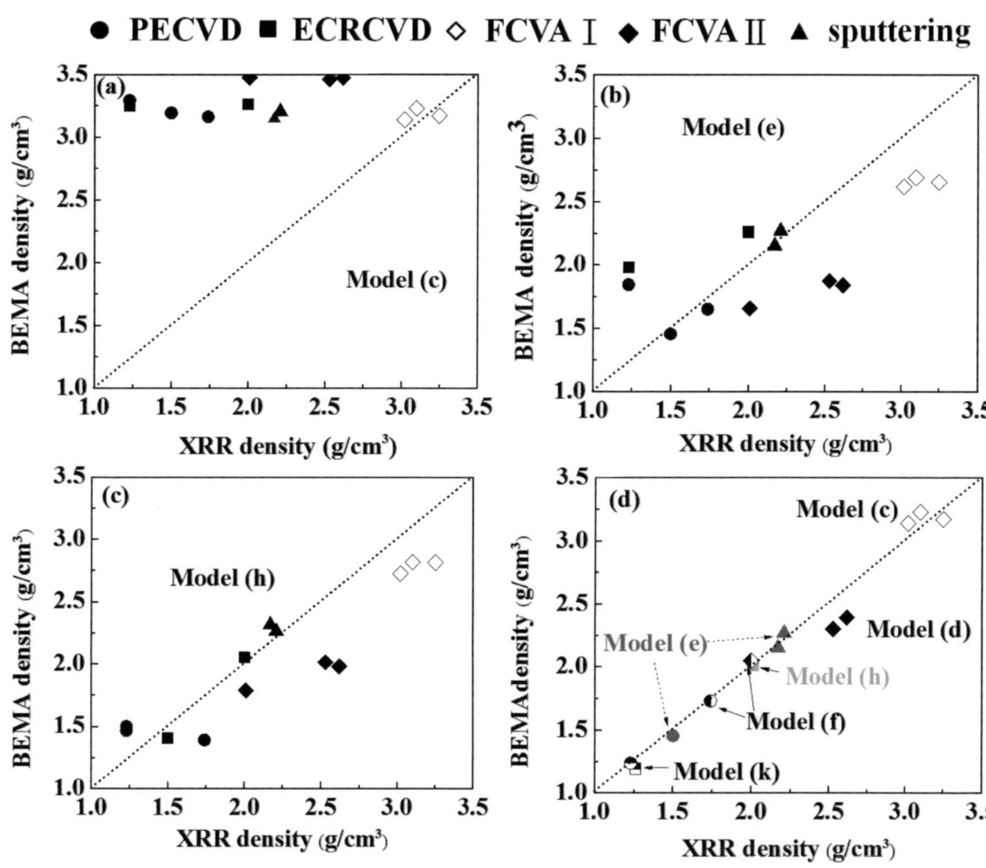


**Figure 8.** The spectra of $n$ (a) and $k$ (b) of DLC films in the photon energy range of 0.6–4.8 eV.

as the photon energy where $n$ has the maximum value in the selected range. The $E_{n\text{-max}}$ for each sample is listed in **Table 3**. **Figure 8(b)** shows the $k$ spectra of the samples A–K in the photon energy range of 0.6–4.8 eV. The $k$ of each sample increases with photon energy. However, the $k$ spectra have no distinct peak characteristic as seen in $n$. We focus on the threshold energy $E_k$ defined as the photon energy where $k$ has the value more than $10^{-4}$. The optical energy gap $E_g$ of DLC films is obtained from the Tauc-Lorentz model fitting of the experimental data.

### 4.2. Discussions

#### *4.2.1. Discussions from experiment part one*

**Figure 9** shows the comparisons of $\rho_{BEMA}$ with $\rho_{XRR}$. Optical models (c) and (d) without void will be useful for the high-density and low-hydrogen content DLC films as follows. **Figure 9(a)** indicates that all the simulation results with the model (c) have provided the $\rho_{BEMA}$ values from 3.0 to 3.5 g/cm$^3$. Only the FCVA samples #01–03 are within this range and have small discrepancies with $\rho_{XRR}$. Model (c) is more suitable for the sample which has the density more than 3.0 g/cm$^3$ than model (d). Samples #04–06 have the hydrogen contents of about 5% which are considered to be the critical value between DLC films of types I and II or between types I and IV, leading to their densities (2.01–2.62 g/cm$^3$) lower than the samples #01–03. Our results suggest that the model (d) using GC instead of PG as the standard material of C($sp^2$) is suitable for these samples. In models (e) and (f), the void is introduced as the free volume. **Figure 9(b)** shows the calculated results using model (e) for the samples #07, #08, and #12 which have the hydrogen contents in the range of 19–36 at.%. Since samples #05 and #11 result in the closed values with those of $\rho_{BEMA}$ by using the model (f), it is also demonstrated that GC plays an important role in the simulation of the low-density DLC films. In models (g) and (h), PE was used as one of the standard materials. The calculated $\rho_{BEMA}$ from the model (g) was lower than the observed $\rho_{XRR}$ when $\rho_{XRR}$ was less than 3.0 g/cm$^3$. Since

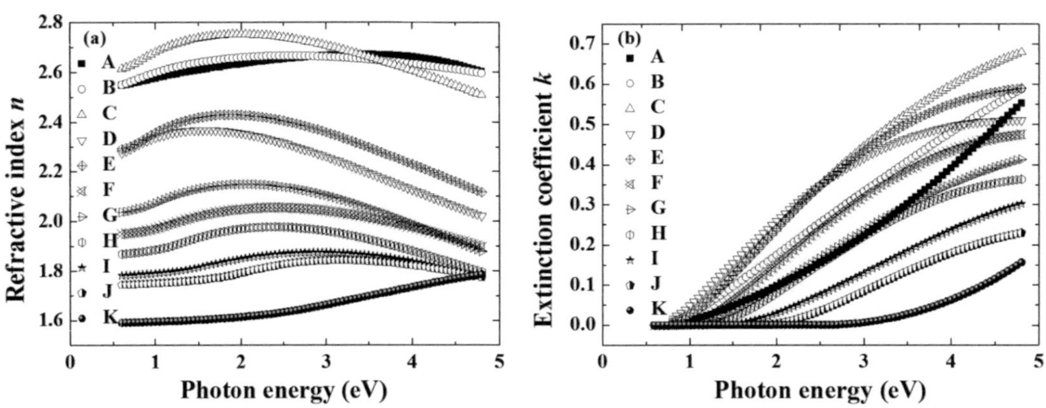


**Figure 9.** Comparison of $\rho_{BEMA}$ and $\rho_{XRR}$, (a) model (c), (b) model (e), (c) model (h), (d) most suitable results in different optical models.

diamond has a similar dielectric function with PE, the low C($sp^3$) (diamond) fraction and the low $\rho_{BEMA}$ will be obtained when diamond and PE are simultaneously included. **Figure 9(c)** shows the results using model (h). Similar to model (g), almost all the samples showed lower $\rho_{BEMA}$ than $\rho_{XRR}$. For samples #07–09, the $\rho_{BEMA}$ values obtained from models (i)–(l) agree with $\rho_{XRR}$. We expect that high-hydrogen content samples, so-called PLC films, can be fully represented by these models. Samples #10 and #13 are appropriately represented by the model (k). Therefore, based on the above analysis and on the combination of the hydrogen content and $\rho_{XRR}$ of all the samples, it can be concluded that each type of DLC film has the suitable optical model. First, it can be expected that the models (i)–(k) are the most suitable for low-density ($\rho_{XRR}$ < 1.4 g/cm$^3$) and high-hydrogen content PLC films (type). The DLC films with the density in the range of 1.4 < $\rho_{XRR}$ < 3.0 g/cm$^3$ corresponding to the types can be simulated by the optical models (e) or (f). Besides, for the DLC films having similar $\rho_{XRR}$ values, which are difficult to distinguish (e.g., samples #04 and #06), optical model (d) may be more suitable than the others. Furthermore, the type DLC films (hydrogen-free ta-C films) having the $\rho_{XRR}$ values of more than 3.0 g/cm$^3$ are suitably simulated by the model (c). However, this chapter has not considered any other factors, such as the density of dangling bonds. Thus, there will be some errors or deviations between the simulation and measurement. **Figure 9(d)** shows the comparison of $\rho_{BEMA}$ and $\rho_{XRR}$. All the BEMA simulations in the present chapter are fully consistent with the measured XRR results. It is therefore indicated that the volume fractions obtained by BEMA theory are the potential parameters for the structural analysis of DLC films.

**Table 4** shows the $sp^3/(sp^3 + sp^2)$ ratios obtained by the fitting of carbon *K*-edge NEXAFS spectra and the volume fraction of each sample. The results show that each type of film has an ideal optical model corresponding to it, and there is no optical model to fit all types of DLC films. Simultaneously, the experimental data prove that the physical deposition method (FCVA or sputtering) is more likely to obtain ta-C and ta-C:H films and the chemical process (ECR or PECVD) more tends to deposit carbon films with higher hydrogen content. This is the fundamental reason why DLC films were able to show quite different physical and chemical properties. The $sp^3/(sp^3 + sp^2)$ ratios obtained from FCVA and sputtering methods have the close values, where the deviations of ratios are around ±0.1. The amount of hydrogen content may be the main reason for the division into the categories. When the hydrogen content exceeds about 25 at.%, PE may be more appropriate as the standard material of C($sp^3$) than diamond because the deviations of $sp^3/(sp^3 + sp^2)$ ratios between BEMA and NEXAFS would be large (see the deviations from samples #07 and #08 to samples #11 and #12) if diamond used. In addition, in the simulation of sample #09, the optical model (h) was used where diamond, PG, and PE as the standard materials were simultaneously introduced. Although PE is artificially introduced as void, it has similar dielectric functions with a diamond which can also represent the C($sp^3$) structure of the DLC films. The $sp^3/(sp^3 + sp^2)$ ratio of sample #09 is 0.49, where the volume fractions of diamond and PE obtained from this simulation analysis result almost in the same values of 33% and 32%. That is to say, even if the PE volume fraction is used, almost the same $sp^3/(sp^3 + sp^2)$ ratio may be obtained. Overall, the comparison between BEMA and NEXAFS

| Sample | Type | Optical model | Volume fraction (%) | | | $sp^3/(sp^3 + sp^2)$ ratio | |
|---|---|---|---|---|---|---|---|
| | | | C($sp^3$) | C($sp^2$) | Void | BEMA | NEXAFS |
| #01 | I | c | 77 | 23 | – | 0.77 | 0.65 |
| #02 | I | c | 70 | 30 | – | 0.70 | 0.58 |
| #03 | I | c | 72 | 28 | – | 0.72 | 0.71 |
| #04 | II or III | d | 39 | 61 | – | 0.39 | 0.49 |
| #05 | II or III | f | 26 | 65 | 9 | 0.29 | 0.38 |
| #06 | II or III | d | 32 | 68 | – | 0.32 | 0.47 |
| #07 | II or IV | e | 34 | 42 | 24 | 0.44 | 0.50 |
| #08 | II or IV | e | 39 | 39 | 22 | 0.50 | 0.57 |
| #09 | IV | h | 33 | 35 | 32 (PE) | 0.49 | 0.29 |
| #10 | VI | k | 68 (PE) | 24 | 8 | 0.74 | 0.42 |
| #11 | IV | f | 21 | 58 | 21 | 0.27 | 0.38 |
| #12 | IV | e | 7 | 54 | 40 | 0.11 | 0.42 |
| #13 | VI | k | 49 (PE) | 34 | 17 | 0.59 | 0.44 |

**Table 4.** Volume fraction and $sp^3/(sp^3 + sp^2)$ ratio of DLC films.

methods shows highly consistent results of the density and $sp^3/(sp^3 + sp^2)$ ratios for the films fabricated by the FCVA and sputtering methods which contain nonhydrogen precursors. When the hydrogen content increases the discrepancy of $sp^3/(sp^3 + sp^2)$ ratios between BEMA and NEXAFS tends to increase and two aspect reasons should be considered. First, in the present chapter, BEMA theory which is used to simulate the experimental data is taken into account only for the shading factors, where each optical model contains the physical combination of different standard materials; the density of dangling bond and the interface effects of different phases and other factors in DLC films have not been considered. Second, the experimental methods of the NEXAFS measurements should be taken into consideration. Usually, the total electron yield (TEY) and PEY modes are used in the measurement of NEXAFS. However, these modes are effective for the samples to have electrical conductivity; the DLC films, especially high-hydrogen content PLC films, have more complex structures close to the insulators, leading to the $sp^3/(sp^3 + sp^2)$ ratio less than the actual value [22]. Therefore, the present method of analysis can be used to distinguish the so-called DLC (type) and PLC films.

#### *4.2.2. Discussion from experiment part two*

**Figure 10(a)** shows the relationship between the present classification scheme based on $n$ and $k$ and the previous one based on the hydrogen contents and the $sp^2/(sp^2 + sp^3)$ ratios with some modifications. According to the later classification, our DLC films are categorized

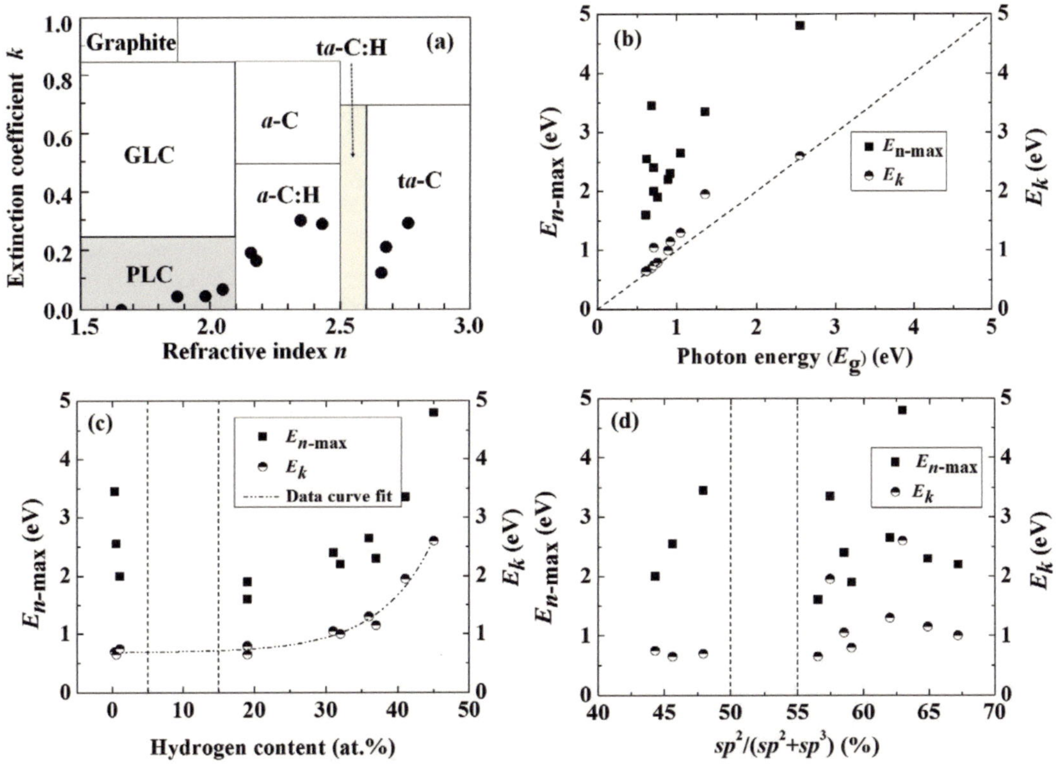


**Figure 10.** The relationship of (a) $n$ and $k$ over the classification of the deposited DLC films modified from past works [11, 12, 34] with wavelength at 550 nm; (b) Tauc-gap $E_g$ and $E_k$ or $E_{n\text{-max}}$; (c) hydrogen contents and $E_k$ or $E_{n\text{-max}}$; (d) $sp^2/(sp^2 + sp^3)$ ratios and $E_k$ or $E_{n\text{-max}}$.

into three types, however, the boundary between PLC and *a*-C:H is still controversial as specified in the range of $n$ = 1.8–2.2 and $k$ = 0.0–0.25 as shown in **Figure 10(a)**. It is necessary to construct a database which is large enough to establish the straightforward method to classify the DLC films in the future. $E_g$ is often used to characterize practical optical properties of DLC materials [30]. **Figure 10(b)** shows $E_k$ and $E_{n\text{-max}}$ as a function of $E_g$ of the present samples. It is found that both $E_k$ and $E_{n\text{-max}}$ are higher photon energy than its $E_g$. In particular, the $E_k$ are distributed in the range of 0.65–2.60 eV and closed to $E_g$. In addition, we found that the hydrogen content and $E_k$ have an exponential relationship and $E_{n\text{-max}}$ also exponentially increases in the hydrogen contents above 15 at.% as shown in **Figure 10(c)**. However, in **Figure 10(d)**, both $E_k$ and $E_{n\text{-max}}$ seem to be an ambiguous correlation with the $sp^2/(sp^2 + sp^3)$ ratios. It might be explained in a way that, in the hydrogenated DLC films, C-H bond plays a crucial role in the cluster formation and its electrical conductivity rather than the $sp^2/(sp^2 + sp^3)$ ratio does [2, 5, 29], even though $E_k$ and $E_{n\text{-max}}$ fairly link with $E_g$ as shown in **Figure 10(b)**. It is noteworthy that a noticeable gap observed in the hydrogen content about 5–15% is possibly corresponding to the gap appeared in the $sp^2/(sp^2 + sp^3)$ ratios between about 50 and 55%. The hydrogen content tends to be a continuous distribution when it exceeds 20% [13, 31], while it seems to be a discontinuous distribution when it under 20% [32, 33]. The existence of the gaps should be an important evidence and root cause that why the DLC films can be classified, and also corresponds to the results of the analysis in **Figure 10(a)**. Thus, the hydrogen content should be the decisive factor affecting the optical properties of DLC films. SE spectral analysis based on $E_k$ and $E_{n\text{-max}}$ can facilitate the assessment of DLC films.

## 5. Conclusions

We analyzed various types of DLC films with different structure and hydrogen contents in the range of 0.3–42% fabricated by FCVA, sputtering, ECRCVD, and PECVD techniques. The structural analysis of a variety of DLC films was performed by using a combination of SE, BEMA, RBS/ERDA, NEXAFS, and XRR. In part one, from the comparisons of the density ($\rho_{XRR}$ and $\rho_{BEMA}$) and $sp^3/(sp^3 + sp^2)$ ratio between BEMA simulation and NEXAFS results, we first established the appropriate optical models of the various types of DLC films and successfully made a detailed structural analysis of the selected DLC films. The results indicated that the BEMA method is effective for the structural analysis of samples. Especially in the so-called DLC films, good agreements are obtained between $\rho_{BEMA}$ and $\rho_{XRR}$ and between the $sp^3/(sp^3 + sp^2)$ ratios from BEMA and NEXAFS methods. Additionally, the hydrogen content plays a nonnegligible role in the structure of all types of DLC films; therefore, the method of the structural analysis of DLC films has to be carefully selected. In part two, through the analysis and comparison of structural analysis and optical constants, we have effectively classified different types of DLC films and achieved a high consistency. The results suggested that the optical constants ($n$ and $k$, $\lambda$ = 550 nm) obtained from SE could be a practical tool to classify the DLC films to some extent. The DLC classification method of optical constants based on SE analysis will be a most practical classification scheme which does not depend on large-scale analysis equipment. Simultaneously, we also found that $E_k$ and $E_{n\text{-max}}$ of the present DLC films exhibit to have exponential dependencies on the hydrogen contents. In particular, the ellipsometrically measured $E_k$ is useful to identify the hydrogen contents of the DLC films in the range from 0 to 50%. In summary, SE as a powerful analytical tool has great potential for simplifying the structural analysis of DLC films and achieving fast and efficient classification. The application of SE on the classification of DLC films need to pour enough attention; we also continue to focus on these research.

## Author details

XiaoLong Zhou* and Hidetoshi Saitoh

*Address all correspondence to: zhou_xiaolong@mst.nagaokaut.ac.jp

Department of Materials Science and Technology, Nagaoka University of Technology, Japan

## References

[1] Bewilogua K, Hofman D. History of diamond-like carbon films – From first experiments to worldwide applications. Surface and Coatings Technology. 2014;**242**:214-225

[2] Zhou XL, Tunmee S, Suzuki T, Phothongkam P, Kanda K, Komatsu K, Kawahara S, Ito H, Saitoh H. Quantitative NEXAFS and solid-state NMR studies of $sp^3/(sp^3 + sp^2)$ ratio in the hydrogenated DLC films. Diamond and Related Materials. 2016;**73**:232-240

[3] Ferrari AC, Libassi A, Tanner BK, Stolojan V, Yuan J, Brown LM, Rodil SE, Kleinsorge B, Robertson J. Density, $sp^3$ fraction, and cross-sectional structure of amorphous carbon films determined by x-ray reflectivity and electron energy-loss spectroscopy. Physical Review B. 2000;**62**(16):11089-11103

[4] Robertson J. Diamond-like amorphous carbon. Materials Science and Engineering R. 2002;**37**(4-6):129-281

[5] Jacob W, Möller W. On the structure of thin hydrocarbon films. Applied Physics Letters. 1993;**63**(13):1771-1773

[6] Gui WG, Lai QB, Zhang L, Wang FM. Quantitative measurements of $sp^3$ content in DLC films with Raman spectroscopy. Surface and Coatings Technology. 2010;**205**(7):1995-1999

[7] Tunmee S, Photongkam P, Euaruksakul C, Takamatsu H, Zhou XL, Wongpaya P, Komatsu K, Kanda K, Ito H, Saitoh H. Investigation of pitting corrosion of diamond-like carbon films using synchrotron-based spectromicroscopy. Journal of Applied Physics. 2016;**120**:195303

[8] Tunmee S, Supruangnet R, Nakajima H, Zhou XL, Arakawa S, Suzuki T, Kanda K, Ito H, Komatsu K, Saitoh H. Study of Synchrotron Radiation Near-Edge X-Ray Absorption Fine-Structure of Amorphous Hydrogenated Carbon Films at Various Thicknesses. Journal of Nanomaterials. 2015;**2015**:1-7

[9] VDI2840. Carbon films basic knowledge, film types and properties. Düsseldorf: Verein Deutscher Ingenieure; 2005

[10] Kim SW, Kim SG. Prospects of DLC coating as environment friendly surface treatment process. Journal of Environmental Sciences. 2011;**23**:S08-S13

[11] Hiramatsu M, Nakamori H, Kogo Y, Sakurai M, Ohtake N, Saitoh H. Correlation between Optical Properties and Hardness of Diamond-Like Carbon Films. Journal of Solid Mechanics and Materials Engineering. 2012;**7**(2):187-198

[12] DIN 50989-1: 2017-04-Draft, Ellipsometry-Part 1: Principles; Text in German and English, Beuth Verlag; 2017

[13] Weiler M, Sattel S, Jung K, Ehrhardt H, Veerasamy VS, Robertson J. Highly tetrahedral, diamond-like amorphous hydrogenated carbon prepared from a plasma beam source. Applied Physics Letters. 1994;**64**:2797-2799

[14] Zhou XL, Arakawa S, Tunmee S, Komatsu K, Kanda K, Ito H, Saitoh H. Structural analysis of amorphous carbon films by BEMA theory based on spectroscopic ellipsometry measurement. Diamond and Related Materials. 2017;**79**:46-59

[15] Niklasson GA, Granqvist CG, Hunderi O. Effective medium models for the optical properties of inhomogeneous materials. Applied Optics. 1981;**20**(1):26-30

[16] Chen ZY, Yu YH, Zhao JP, Wang X, Liu XH, Shi TS. Determination of the sp3/sp2 ratio in tetrahedral amorphous carbon films by effective medium approximation. Journal of Applied Physics. 1998;**83**(3):1281-1285

[17] Zhou XL, Suzuki T, Nakajima H, Komatsu K, Kanda K, Ito H, Saitoh H. Structural analysis of amorphous carbon films by spectroscopic ellipsometry, RBS/ERDA, and NEXAFS. Applied Physics Letters. 2017;**110**:201902

[18] Safaie P, Eshaghi A, Bakshi SR. Optical properties of oxygen doped diamond-like carbon thin films. Journal of Alloys and Compounds. 2016;**672**:426-432

[19] Papadopoulos AD, Anastassakis E. Optical properties of diamond. Physical Review B. 1991;**43**(6):5090-5097

[20] Tajir D, Tougaard S. Electronic and optical properties of selected polymers studied by reflection electron energy loss spectroscopy. Journal of Applied Physics. 2012;**111**(5):054101

[21] Guo WS, Wong SP, Yu YH. Spectroscopic ellipsometry characterization of diamond-like carbon films formed by filtered arc deposition. Nuclear Instruments and Methods in Physics Research Section B. 2000;**169**(1-4):54-58

[22] Battie Y, Broch L, En Naciri A, Lauret J-S, Guézo M, Loiseau A. Diameter dependence of the optoelectronic properties of single walled carbon nanotubes determined by ellipsometry. Carbon. 2015;**83**:32-39

[23] Fujiwara H. Spectroscopic Ellipsometry: Principles and Applications. London: John Wiley & Sons Ltd; 2007

[24] Palik ED. Handbook of Optical Constants of Solid. London: Elsevier; 1997

[25] Williams MW, Arakawa ET. Optical properties of glassy carbon from 0 to 82 eV. Journal of Applied Physics. 1972;**43**(8):3460-3463

[26] Iwaki M, Terashima K. Change in atomic density of glassy carbon by Na ion implantation. Surface and Coatings Technology. 2000;**128-129**:429-433

[27] Kawai T, Keller A. On the density of polyethylene single crystals. Philosophical Magazine. 2006;**8**(91):1203-1210

[28] Sauro JP, Bindell J, Wainfan N. Some Observations on the Interference Fringes Formed by X Rays Scattered from Thin Films. Physics Review. 1966;**143**:439-443

[29] Tamor MA, Haire JA, Wu CH, Hass KC. Correlation of the optical gaps and Raman spectra of hydrogenated amorphous carbon films. Applied Physics Letters. 1989;**54**:123-125

[30] Cho G, Yen BK, Klug CA. Structural characterization of sputtered hydrogenated amorphous carbon films by solid state nuclear magnetic resonance. Journal of Applied Physics. 2008;**104**:013531

[31] Tillmann W, Hoffmann F, Momeni S, Heller R. Hydrogen quantification of magnetron sputtered hydrogenated amorphous carbon (a-C:H) coatings produced at various bias voltages and their tribological behavior under different humidity levels. Surface and Coatings Technology. 2011;**206**:1705-1710

[32] Kaplan S, Jansen F, Machonkin M. Characterization of amorphous carbon-hydrogen films by solid-state nuclear magnetic resonance. Applied Physics Letters. 1985;**47**:750-753

[33] Ito H, Yamamoto K, Masuko M. Thermal stability of UBM sputtered DLC coatings with various hydrogen contents. Thin Solid Films. 2008;**517**:1115-1119

[34] Schneider D, Schwarz T. A photoacoustic method for characterising thin films. Surface and Coatings Technology. 1997;**91**:136-146